Lab-on-a-Chip

Techniques, Circuits, and Biomedical Applications

For a complete listing of titles in the
Artech House Integrated Microsystems Series,
turn to the back of this book.

Lab-on-a-Chip

Techniques, Circuits, and Biomedical Applications

Yehya H. Ghallab

Wael Badawy

ARTECH
HOUSE

BOSTON | LONDON
artechhouse.com

Library of Congress Cataloging-in-Publication Data
A catalog record for this book is available from the U.S. Library of Congress.

British Library Cataloguing in Publication Data
A catalogue record for this book is available from the British Library.

Cover design by Vicki Kane

ISBN 13: 978-1-59693-418-4

685 Canton Street
Norwood, MA 02062

10 9 8 7 6 5 4 3 2 1

To the memory of my father Dr. Hassan Ghallab who encouraged me in my research and who always supported my efforts, to my mother, brothers, and sisters, to my wife, Rehab, and my children, Ahmed and Rola

—Yehya H. Ghallab

Contents

Preface

Lab-on-a-chip technology is exciting the interest of scientists in many fields. This technology can be used to synthesize chemicals, biological cells, cancer cells, and DNA efficiently and economically. It also has applications in life science and health care. The merging between electronics used in biology and micro- and nanoscience will clearly be the technological revolution of the future.

Lab-on-a-chip promises many advantages including improved performance, portability, reliability, and cost reduction. Lab-on-a-chip combines many parts and components. The functions of the lab-on-a-chip fall in four major categories: actuators, sensors, read out circuits, and microfludic channels.

In support of this research, this book discusses and reviews the state of the art of the lab-on-a-chip research and its applications at the cell level. Also, this book combines fundamental and basic information about the different components of the lab-on-a-chip. Moreover, this book covers some readout circuits, which are applicable for integration into the lab-on-a-chip applications.

Acknowledgments

The authors express their deep gratitude to Dr. Graham Jullien with the Electrical and Computer Engineering Department, University of Calgary, Calgary, AB, Canada, for his valuable support. The authors are also grateful to the Natural Science and Engineering Research Council of Canada (NSERC) for funding the research and to CMC Microsystem for the support provided in fabricating the microchip used in this book.

1

Introduction to Lab-on-a-Chip

Lab-on-a-chip is a device that integrates one or several laboratory functions on a single chip of only millimeters to a few square centimeters in size [1–3]. Lab-on-a-chip deals with extremely small fluid volumes (it can be less than picoliters). Lab-on-a-chip is often indicated by micro total analysis systems (μTAS) [4–6]. However, lab-on-a-chip generally indicates the scaling of single or multiple lab processes down to chip format, whereas μTAS is dedicated to the integration of the total sequence of lab processes to perform chemical analysis [7, 8]. In other words, lab-on-a-chip showed that μTAS technologies were more widely applicable than only for analysis purposes.

Recently, companies and applied research groups have become interested in different lab-on-a-chip applications such as chemical analysis, environmental monitoring, and medical diagnostics. However, the application of lab-on-a-chip is still novel and modest [9]. Lab-on-a-chip applications include synthetic chemistry (e.g., rapid screening and microreactors for pharmaceutics). Also, research in lab-on-a-chip is expected to extend towards downscaling of fluid handling structures by using nanotechnology [10, 11]. Submicrometer and nanosized channels, DNA labyrinths, single cell detection, and nanosensors become feasible, which allow new ways of interaction with biological species and large molecules [12]. Automated patch clamp chips are a commercially successful example of lab-on-a-chip in life science, which allow for drastically increased throughput for drug screening in the pharmaceutical industry [13].

1.1 History

Miniaturization and integration are the two main keywords in device engineering research, and these two keywords are also the main characteristics of lab-on-a-chip technology. The rapid advance of integrated circuit (IC) technology is the main reason for this big trend. In 1958, the discovery of microtechnology opened the door for integrated semiconductor structures for microelectronic chips [16]. Figure 1.1 shows the first IC. In 1966, these lithography-based technologies were applied in pressure sensor manufacturing [17]. As a result of further development of the IC technology, complementary metal oxide semiconductor (CMOS) compatibility limited processes appeared and were used. Furthermore, a toolbox became available to create micrometer- or submicrometer-sized mechanical structures in silicon wafers. New technologies, such as microelectromechanical systems (MEMS), also had started to become used [18].

Fluid-handling devices were developed after other types of sensors, such as pressure and airbag sensors. These fluid-handling devices are channels (capillary connections), mixers, valves, pumps, and dosing devices [19, 20]. In 1975, S. C. Terry from Stanford University presented a gas chromatographic air analyzer fabricated on a silicon wafer using integrated circuit technology [21]. It was the first μTAS that used techniques borrowed from microelectronics to fabricate a system for chemical analysis. However, lab-on-a-chip research started to seriously grow in the late 1980s and early 1990s as a few research groups in Europe and North America developed micropumps, flow sensors, and the concepts for integrated fluid treatments for analysis systems. These research groups, using lab-on-a-chip concepts, demonstrated that integration of pretreatment steps, usually done at lab scale, could extend the simple sensor functionality towards

Figure 1.1 First integrated circuit chip. (Courtesy of Texas Instruments, Inc.)

a complete laboratory analysis, such as additional cleaning and separation steps. In 1990, μTAS technologies provided interesting tooling for genomics applications, like capillary electrophoresis and DNA microarrays [4]. This resulted in a big boost in research and commercial interest. Also, research support and interest in the field of portable bio/chemical warfare agent detection systems came from the military. The term "lab-on-a-chip" was introduced as a result of this boost, which was not only limited to integration of lab processes for analysis but also the characteristic possibilities of individual components and the application to nonanalysis lab processes [9].

1.2 Parts and Components of Lab-on-a-Chip

Lab-on-a-chip contains many parts and components. The function of the lab-on-a-chip can be fall into three major categories: actuators, sensors, and readout circuits. In the actuation applications, lab-on-a-chip generates either mechanical or electrical forces in fluid or objects via electromagnetic interactions [14]. As a sensor, lab-on-a-chip is used to measure electrical, optical, magnetic, or thermal properties of the target samples. As a readout circuit, lab-on-a-chip is used to amplify, reduce noise, and condition the output signal. It also provides an interface between the lab-on-a-chip and computers [15]. Figures 1.2–1.4 show the components of lab-on-a-chip, which include a chip and a chamber that when combined form a lab-on-a-chip.

1.2.1 Electric and Magnetic Actuators

As an actuator, lab-on-a-chip exerts forces on small objects such as DNA, which are biological cells suspended in fluids. Electric or magnetic fields can be used to create forces and control the motion of objects in fluids. Microelectrodes are implemented to generate an electric field on the lab-on-a-chip's surface. This field interacts directly with the target samples and the surrounding media. For example, in dielectrophoresis applications, cells exploit the difference in polarization between the object and media under an ac electric field to trap and move the objects (see Chapter 3).

For magnetic actuation, microelectromagnets are implemented on the lab-on-a-chip's surface to generate a local magnetic field. Biological cells can be manipulated in the chip's surface. The dimension of the electrodes is in the range of micrometers, which allows the lab-on-a-chip's electrodes to be arranged in an array form. Thus, parallel manipulation of target samples can be performed with high manipulation accuracy and throughput.

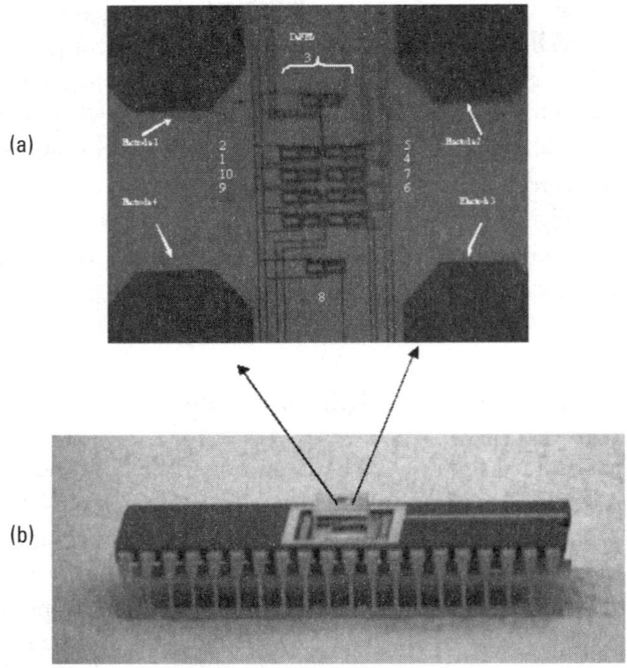

Figure 1.2 (a) Close-up of an integrated circuit chip, which shows the different parts such as electrodes and sensors. (b) The integrated circuit chip (0.7 × 0.7 mm) and a glass spacer.

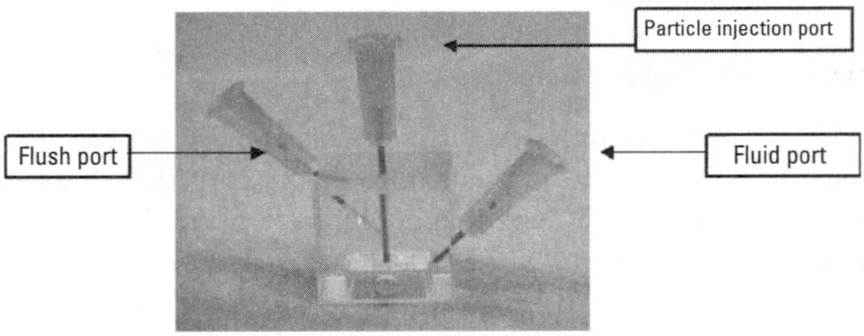

Figure 1.3 The microfluidic chamber.

1.2.2 Electrical Sensors

Microelectrodes in a lab-on-a-chip's surface can form high sensitive electric sensors. Biological cells can be capacitively coupled with microelectrodes and their bioelectrical activities can be measured noninvasively. By implementing a large array of microelectrodes with multiplexing circuits, the propagation of the electrical signals in biological tissues can be monitored and provide valuable information on intercellular communication. CMOS-based microelectrodes have been used to monitor the electrical activity of cultured neurons (see Chapter 7).

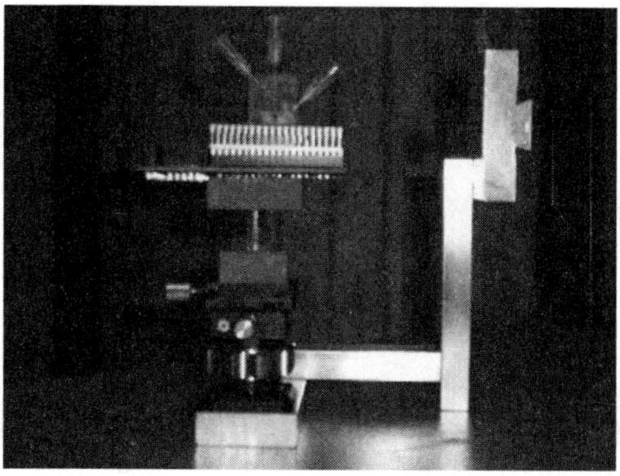

Figure 1.4 The microfluidic chamber above the IC chip, which is installed on the top of the X, Y, Z table.

1.2.3 Thermal Sensors

Precision temperature sensors can be easily integrated in the lab-on-a-chip along with other sensors and actuation components. These sensors provide accurate temperature measurements. In lab-on-a-chip applications, measuring the surface temperature is important for the biocompatibility process as lab-on-a-chip generates heat during its operation. To enhance the biocompatibility of the system, the output of the temperature sensors is coupled with coolers or heaters to maintain the surface temperature regulations.

1.2.4 Optical Sensors

Optical imaging is the most sensitive detection method in biological and medical applications. However, it is the weakest in device miniaturization; optical detection involves bulky equipment such as a light source and microscopes. In lab-on-a-chip, charge-coupled device (CCD) and CMOS imagers are combined with microfluidic systems to image biological samples without relying on intermediary or bulky optical components.

1.2.5 Microfluidic Chambers

Packaging of the lab-on-a-chip is vital. For successful packaging, knowledge of materials, devices, reliability, and a good understanding of the limits of current packaging technology are required. The functions of packaging are to protect the devices from the environment and vice versa. Protection of the device includes electrical insulation and passivation of the device from the penetrating of moisture and ions. On the other hand, it is also necessary to protect the environment from the device materials and the device's operation. Thus, no undesirable interaction with the environment occurs. An example of the microfluidic chamber is shown in Figure 1.3. This chamber was machined out of Lucite plastic. To complete the chamber assembly, 4-mm needles were glued (epoxy) onto the chamber primary fluid ports (see Figure 1.3). A complete system is shown in Figure 1.4. A glass spacer (3 mm thick) was bonded into the chip package (40 PIN DIP IC carrier) well using adhesive glue. The IC chip (Figure 1.2) was subsequently bonded on top of the spacer, as shown in Figure 1.2. Electrical connections (wire bonds) were made from the IC chip pads to the chip package bond pads. Finally, the chamber was mounted above the IC chip. A complete system including the elevator (i.e., X, Y, Z table), which raises the system to the same level as the horizontal microscope, is shown in Figure 1.5.

1.3 Applications of Lab-on-a-Chip

Due to rapid development and a growing interest in medicine, biotechnology, drug discovery, and environmental monitoring, the world has become increasingly dependent on chemical analysis. Previously, chemical analyses have been performed in central laboratories because they require skilled personnel and specific equipment. However, the trend is to make chemical analysis much easier and thus possible for a user who doesn't have experience with chemical equipment. Some examples are pregnancy tests, blood glucose concentration tests for diabetic patients, and analysis for water samples. These tests can be obtained on store shelves and can be used at home by individuals with no special training in chemistry. This trend of simplification of chemical analysis is expected to continue. To achieve this goal, analytical equipment needs to be smaller and thus portable, easier to operate, and reliable. The results of the chemical tests need to be easy for the user to interpret. The concept of lab-on-a-chip builds on performing all of the necessary steps required for a chemical analysis on a miniaturized

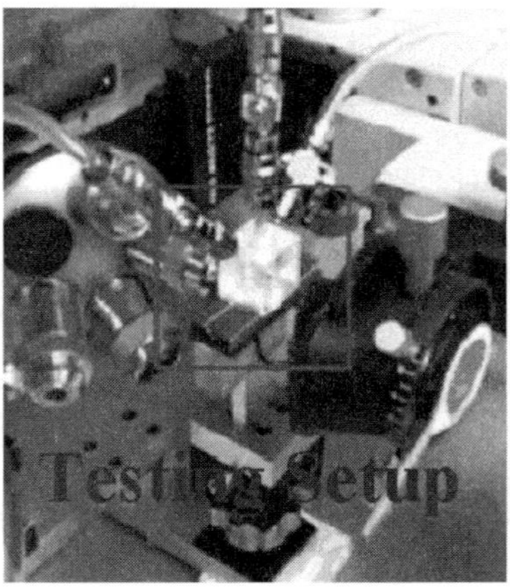

Figure 1.5 A complete system.

system and therefore offers portability. Battery-operated analytical equipment opens up the possibility of performing chemical analysis with very low power independently on the power network. This because of the miniaturized components of the lab-on-a-chip can work with very low power consumption. A fully automated system, which includes the automation of the entire chemical analysis process and data processing, is also a part of the lab-on-a-chip concept. Lab-on-a-chip can be used as a black box where the user needs only to push a start button to perform the analysis and retrieve the results. Microfabrication allows the reproduction of the same lab-on-a-chip with the same specifications. One of the most important component of the lab-on-a-chip is the chip in which a sample of microliters and reagents are moved around with very high accuracy. Conventionally, chemical analyses are performed by mixing milliliters of samples and reagents in a test tubes and analyzing the product in an instrument, such as spectrophotometer. Lab-on-a-chip offers a significant decrease in the cost especially when the samples and reagents are very expensive. The decrease in the cost is achieved by dramatically reducing the volume of samples and reagents that are needed to perform the chemical analysis.

Lab-on-a-chip has many applications in the medical and biological fields. In the past decade, it has started to be used in many other fields, such as environmental applications. In the medical field, it can be used in real-time polymerase chain reaction (PCR) applications to detect bacteria, viruses, and cancers [22, 23]. Lab-on-a-chip is also used in biochemical assays [1]. In immunoassay applications, lab-on-a-chip is used to detect bacteria, viruses, and cancers based on antigen-antibody reactions [2]. Another area of applications is the dielectrophoresis applications where lab-on-a-chip is used to detect cancer cells and bacteria [24]. Lab-on-a-chip can be used in blood sample preparation to crack cells and extract DNA [25, 26]. It is also used in cellular lab-on-a-chip for single-cell analysis and ion channel screening [27–29].

Lab-on-a-chip technology may soon become an important part of efforts to improve global health [30], particularly through the development of point-of-care testing devices. In countries with few healthcare resources, infectious diseases that would be treatable in a developed nation are often deadly. In some cases, poor healthcare clinics have the drugs to treat a certain illness but lack the diagnostic tools to identify patients who should receive the drugs. Many researchers believe that lab-on-a-chip technology may be the key to powerful new diagnostic instruments. The goal of these researchers is to create microfluidic chips that will allow healthcare providers in poorly equipped clinics to perform diagnostic tests such as immunoassays and nucleic acid assays with no laboratory support.

One active area of lab-on-a-chip research involves ways to diagnose and manage HIV infections. Around 40 million people are infected with HIV in the world today, yet only 1.3 million of these people receive antiretroviral treatment. Around 90% of people with HIV have never been tested for the disease [31].

Measuring the number of CD4+ T lymphocytes in a person's blood is an accurate way to determine if a person has HIV and to track the progress of an HIV infection. At the moment, flow cytometry is the gold standard for obtaining CD4 counts, but flow cytometry is a complicated technique that is not available in most developing areas because it requires trained technicians and expensive equipment.

Lab on-a-chip has applications in the environment field, such as real-time monitoring. Real-time monitoring is very important in many applications (e.g., continuous analysis of ammonium in wastewater, which is better than taking measurements only two or three times a day).

1.4 Advantages and Disadvantages of Lab-on-a-Chip

Lab-on-a-chip brings many advantages and promises to science and industry. These advantages are:

- *Increased spatial resolution:* Lab-on-a-chip provides a way to perform experiments with high spatial resolution. This is because sensors and actuators in the integrated chip (IC) have small footprints, which increase the spatial resolution. Also, by using new and updated technologies, such as complementary metal oxide semiconductor (CMOS) technology to manufacture lab-on-a-chip, lab-on-a-chip improves the accuracy of experiments. Lab-on-a-chip can be designed to operate in the gigahertz range for high-speed measurement, which enhances the spatial resolution of the measurement.
- *Automated measurement:* In lab-on-a-chip, actuators and sensors are used in an array format. Consequently, lab-on-a-chip allows users to perform measurement in a parallel fashion for high throughput operation [1–3]. The measurement sequence can be automated by programming the controllers of the lab-on-a-chip, which also minimizes the manmade errors.
- *Robustness:* Durability and reproducibility are significant merits of the lab-on-a-chip. Lab-on-a-chip is manufactured in a commercial foundry with many sophisticated process controls. Thus, the chip-to-chip variation is minimal, which maintains the consistency of the chip performance.
- *User-friendly interface:* Lab-on-a-chip combines different technologies; one of these technologies is the IC technology, which allows a simple interface to the external inputs. Also, it provides different user-friendly interfaces. Thus, lab-on-a-chip users can use the same user-friendly interface in lab-on-a-chip applications.

- *Portability and disposability:* Lab-on-a-chip is a combination of both the IC and microfluidic systems. The IC includes many parts, such as sensing, actuation, and detection parts. Also IC includes the electronic circuits, which control different functions. Lab-on-a-chip is a cost-effective portable unit; it can be mass produced using the conventional IC foundries. The microfluidic systems can be fabricated at a low cost using polymers. Therefore, lab-on-a-chip can be cheap and disposable, which is very attractive in biomedical applications.
- *New avenues and opportunities for IC industry:* The IC industry can take good and new opportunities from lab-on-a-chip. New markets can be opened as well. Also, lab-on-a-chip will benefit from newly developed technology.

However, lab-on-a-chip still has many challenges and disadvantages. They are:

- *Novelty:* Lab-on-a-chip is a novel technology and therefore not yet fully developed.
- *Physical and chemical effects:* In lab-on-a-chip fabrication, which is a small-scale industry, there are many effects that become more dominant such as capillary forces, surface roughness, and chemical interactions of construction materials on reaction processes. This can sometimes make processes in lab-on-a-chip more complex than in conventional lab equipment.
- *Signal-to-noise ratio:* Detection principles may not always scale down in a positive way, leading to low signal-to-noise ratios.
- *Accuracy and precision:* although the absolute geometric accuracies and precision in microfabrication are high, they are often rather poor in a relative way compared to precision engineering, for instance.

References

[1] Oosterbroek, E., and A. van den Berg, (eds.), *Lab-on-a-Chip: Miniaturized Systems for (Bio) Chemical Analysis and Synthesis*, 2nd ed., New York: Elsevier Science, 2003.

[2] Klank, G., and E. Telleman, *Microsystem Engineering of Lab-on-a-Chip Devices*, New York: John Wiley & Sons, 2004.

[3] Herold, K., and A. Rasooly, (eds.), *Lab-on-a-Chip Technology: Fabrication and Microfluidics*, Norfolk, U.K.: Caister Academic Press, 2009.

[4] Ohori, T., et al., "Three-Way Microvalve for Blood Flow Control in Medical Micro Total Analysis Systems (μTAS)," *IEEE 10th Annual International Workshop on Micro Electro Mechanical Systems (MEMS '97)*, 1997, pp. 333–337.

[5] Yoshida, H., T. Sasa, and E. Maruyama, "Co-Evolution Process Between Basic Research and Applied Research: A Case Study of μ-TAS Projects," *Portland International Conference on Management of Engineering & Technology (PICMET 2009)*, 2009, pp. 1425–1431.

[6] Johansson, L., et al., "Temperature and Trapping Characterization of an Acoustic Lateral Trap for μ-TAS," *International Solid-State Sensors, Actuators, and Microsystems Conference (TRANSDUCERS 2007)*, 2007, pp. 739–742.

[7] Chia-Fu, C., and F. Zenhausern, "Electrodeless Dielectrophoresis for Micro Total Analysis Systems," *IEEE Engineering in Medicine and Biology Magazine*, Vol. 22, No. 6, 2003, pp. 62–67.

[8] Hartley, L., K. Kaler, and O. Yadid-Pecht, "Hybrid Integration of an Active Pixel Sensor and Microfluidics for Cytometry on a Chip," *IEEE Trans. on Circuits and Systems I: Regular Papers*, 2007, pp. 99–110.

[9] Ghallab, Y. H., and W. Badawy, "Sensing Methods of Dielectrophorieses from Bulky Instruments to Lab-on-a-Chip," *IEEE Circuits and Systems Magazine*, Vol. 4, No. 3, 2004, pp. 5–15.

[10] Fortina, P., S. Surrey, and L. Kricka, "Molecular Diagnostics: Hurdles for Clinical Implementation," *Trends in Molecular Medicine*, Vol. 8, 2002, pp. 264–266.

[11] Jain, K. K., "Pharmacogenomics," *Third Annual Conference on Lab-on-a-Chip and Microarrays*, Vol. 2, Zurich, Switzerland, 2001, pp. 73–77.

[12] Kricka, L., "Microchips, Microarrays, Biochips and Nanochip: Personal Laboratories for the 21st Century," *Clinica Chimica Acta*, Vol. 307, 2001, pp. 219–223.

[13] Medoro, G., et al., "A Lab-on-a-Chip for Cell Separation Based on the Moving Cages Approach," *16th European Conference on Solid State Transducers*, Prague, Czech Republic, 2002.

[14] Medoro, G., et al., "CMOS-Only Sensors and Manipulation for Microorganisms," *Proceedings of IEDM*, 2000, pp. 415–418.

[15] Medoro, G., et al., "A Lab-on-a-Chip for Cell Detection and Manipulation," *IEEE Sensors Journal*, Vol. 3, No. 3, 2003, pp. 317–325.

[16] Kilby, J. S., "Invention of the Integrated Circuit," *IEEE Trans. on Electron Devices*, Vol. ED-23, 1976, pp. 648–654.

[17] Moore, G. E., "Cramming More Components onto Integrated Circuits," *Electronics*, Vol. 38, 1965, pp. 114–117.

[18] Bustillo, J. M., R. T. Howe, and R. S. Muller, "Surface Micromachining for Microelectromechanical Systems," *Proceedings of the IEEE*, Vol. 86, 1998, pp. 1552–1574.

[19] Manz, A., N. Graber, and H. M. Widmer, "Miniaturized Total Chemical Analysis Systems: A Novel Concept for Chemical Sensing," Vol. 1, 1990, pp. 244–248.

[20] Terry, S.C., G. H. Jerman, and J. B. Angell, "A Gas Chromatographic Air Analyzer Fabricated on a Silicon Wafer," *IEEE Trans. on Electron Devices*, Vol. ED-26, 1979, pp. 1880–1886.

[21] Bhushan, A., et al., "Fabrication and Preliminary Results for LiGA Fabricated Nickel Micro Gas Chromatograph Columns," *Journal of Microelectromechanical Systems*, Vol. 16, No. 2, 2007, pp. 383–393.

[22] Fukuba, T., T. Naganuma, and T. Fujii, "Microfabricated Flow-Through PCR Device for Underwater Microbiological Study," *Proceedings of the 2002 International Symposium on Underwater Technology*, 2002, pp. 101–105.

[23] Mock, J. E., M. H. Gordon, and W. W. Bradley, "Demonstration of a Flow-Through Micro-PCR in an Annular Pyrex Channel," *IEEE Region 5 Technical Conference*, 2007, pp. 415–418.

[24] Pohl, H. A., *Dielectrophoresis*, Cambridge, U.K.: Cambridge University Press, 1978.

[25] Washizu, M., and O. Kurosawa, "Electrostatic Manipulation of DNA in Microfabricated Structures," *IEEE Trans. on Industry Applications*, Vol. 26, No. 6, 1990, pp. 1165–1172.

[26] Casanella, R., et al., "Aggregation Profile Characterization in Dielectrophoretic Structures Using Bacteria and Submicron Latex Particles," *IEE Proceedings on Nanobiotechnology*, Vol. 150, 2003, pp. 70–74.

[27] Ghallab, Y. H., and W. Badawy, "A CMOS Lab-on-a-Chip for Biomedical Applications," *International Symposium on Circuit and System (ISCAS)*, Japan, May 23, 2005.

[28] Medoro, G., et al., "A Lab-on-a-Chip for Cell Detection and Manipulation," *Proceedings of IEEE Sensors*, Vol. 1, Orlando, FL, 2002, pp.472–475.

[29] Romani, A., et al., "Capacitive Sensor Array for Localization of Bioparticles in CMOS Lab-on-a-Chip," *IEEE International Solid-State Circuits Conference (ISSCC 04)*, 2004, pp. 456–457.

[30] Yager, P., et al., "Microfluidic Diagnostic Technologies for Global Public Health," *Nature*, Vol. 442, No. 7101, 2006, pp. 412–418.

[31] Craig, I., and X. Xiaohua, "Can HIV/AIDS Be Controlled? Applying Control Engineering Concepts Outside Traditional Fields," *IEEE Magazine for Control Systems*, Vol. 25, No. 1, 2005, pp. 80–83.

2

Cell Structure, Properties, and Models

The cell is the smallest unit of an organism that is classified as living, and is often called the building block of life. Based on modern cell theory, the cell is the fundamental unit of structure and function in living things—all cells come from preexisting cells by division. Also, energy flow (metabolism and biochemistry) occurs within cells [1–3]. Cells contain hereditary information (DNA), which is passed from cell to cell during cell division. All cells are basically the same in chemical composition [1, 2, 4]; all known living things are made up of cells.

The cell is the structural and functional unit of all known living organisms, which includes human, animal, plant, fungus, or micro-organism. An organism may either be unicellular (single-celled) or be composed of (as in humans) many billions of cells grouped into specialized tissues and organs. The term multicellular describes any organism made up of more than one cell [5–7].

In this chapter, the focus will be on two main cell related topics, they are particle electromechanics and electrogenic cells. These topics will help the reader better understand the coming chapters of this book.

2.1 Cell Structure

Cells can be subdivided into two categories:

1. *Prokaryotes*, which lacks a nucleus and other membrane-bound organelles, although they contain ribosomes [8]. Bacteria and archaea are two divisions of prokaryotes.

2. *Eukaryotes*, which have distinct nuclei and membrane-bound organelles (i.e., mitochondria, chloroplasts, lysosomes, rough and smooth endoplasmic reticulum, vacuoles). They possess organized chromosomes, which store genetic material [9].

2.1.1 Prokaryotic Cells

The prokaryote cell is simpler than a eukaryote cell, lacking a nucleus and most of the other organelles of eukaryotes. There are two kinds of prokaryotes: bacteria and archae [10–14]. Figure 2.1 shows the structure of a prokaryotic cell. This figure shows that prokaryotic cell has three architectural regions:

1. *Outside region:* Flagella and pili project from the cell's surface. These are structures (not present in all prokaryotes) made of proteins that facilitate movement and communication between cells.
2. *Cell envelope:* Generally consisting of a cell wall covering a plasma membrane. The envelope gives rigidity to the cell and separates the interior of the cell from its environment, serving as a protective filter. Most prokaryotes have a cell wall; there are exceptions such as mycoplasma (bacteria) and thermoplasma (archaea). The cell wall consists of peptidoglycan in bacteria, and acts as an additional barrier against exterior

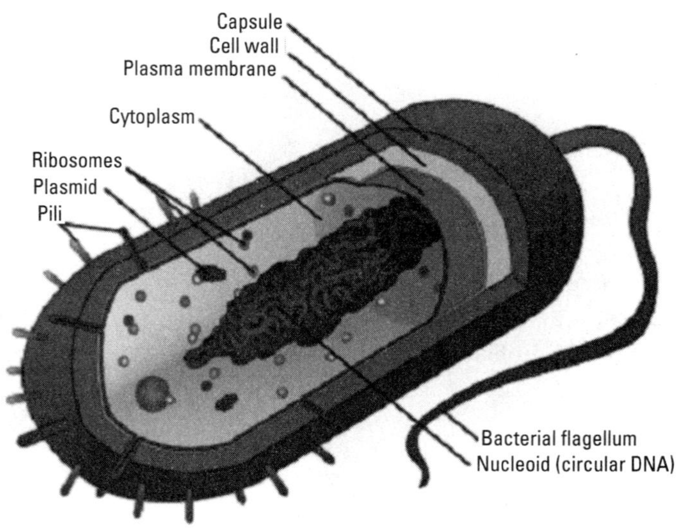

Figure 2.1 Prokaryotic cells [10].

forces. It also prevents the cell from expanding and finally bursting (cytolysis) from osmotic pressure against a hypotonic environment.

3. *Inside region:* The cytoplasmic region contains the cell genome (DNA), ribosomes, and various sorts of inclusions. A prokaryotic chromosome is usually a circular molecule. The DNA is condensed in a nucleoid; however, it does not form a nucleus. Prokaryotes can carry extrachromosomal DNA elements called plasmids, which are usually circular. Plasmids enable additional functions, such as antibiotic resistance.

2.1.2 Eukaryotic Cells

Eukaryotic cells are about 10 times the size of a typical prokaryote and can be as much as 1,000 times greater in volume [15–17]. Eukaryotic cells contain membrane-bound compartments in which specific metabolic activities take place; this is the major difference between prokaryotes and eukaryotes. Other differences are also presented between prokaryotes and eukaryotes, such as the presence of a cell nucleus, which is the most important among these differences and a membrane-delineated compartment that houses the eukaryotic cell's DNA. Another difference is that the plasma membrane resembles that of prokaryotes in function, with minor differences in the setup. Cell walls may or may not be present [18, 19]. The eukaryotic DNA is organized in one or more linear molecules, called chromosomes, which are associated with histone proteins. All chromosomal DNA is stored in the cell nucleus, separated from the cytoplasm by a membrane. Some eukaryotic organelles such as mitochondria also contain some DNA [20]. Many eukaryotic cells are ciliated with primary cilia. Primary cilia play important roles in chemosensation, mechanosensation, and thermosensation. Cilia may thus be "viewed as sensory cellular antennae that coordinate a large number of cellular signaling pathways, sometimes coupling the signaling to ciliary motility or alternatively to cell division and differentiation" [21]. Eukaryotes can move using motile cilia or flagella. The flagella are more complex than those of prokaryotes. Figure 2.2 shows the structure of an animal cell as an example of eukaryotic cell.

2.1.3 Cell Components

2.1.3.1 Cell Membrane

All cells, whether prokaryotic or eukaryotic, have a membrane that envelops the cell, separates its interior from its environment, regulates what moves in and out (selectively permeable), and maintains the electric potential of the cell [1]. Inside the membrane, a salty cytoplasm takes up most of the cell volume. The plasma membrane in plants and prokaryotes is usually covered by a cell wall [2,

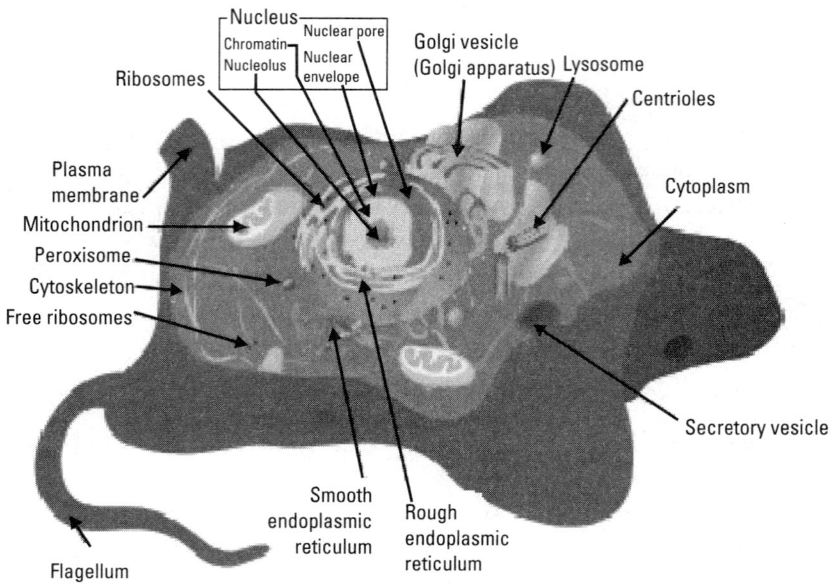

Figure 2.2 Eukaryotic cell (animal) [11].

3]. This membrane serves to separate and protect a cell from its surrounding environment and is made mostly from a double layer of lipids (hydrophobic fat-like molecules) and hydrophilic phosphorus molecules. Hence, the layer is called a phospholipid bilayer. Embedded within this membrane is a variety of protein molecules that act as channels and pumps moving different molecules into and out of the cell. The membrane is said to be "semipermeable," in that it can either let a substance (molecule or ion) pass through freely, pass through to a limited extent, or not pass through at all. Cell surface membranes also contain receptor proteins that allow cells to detect external signaling molecules such as hormones [1].

2.1.3.2 Cytoskeleton

The cytoskeleton's main functions are:

- Organize and maintain the cell's shape;
- Anchor organelles in place;
- Help during endocytosis, the uptake of external materials by a cell;
- Help during cytokinesis, the separation of divided cells after cell division;
- Move parts of the cell in processes of growth and mobility [20].

The eukaryotic cytoskeleton is composed of microfilaments, intermediate filaments, and microtubules. There are a great number of proteins associated with them, each controlling a cell's structure by directing, bundling, and aligning filaments [21]. The prokaryotic cytoskeleton is less well studied, but is involved in the maintenance of cell shape, polarity, and cytokinesis [22].

2.1.3.3 Genetic Material

Two different kinds of genetic material exist: deoxyribonucleic acid (DNA) and ribonucleic acid (RNA) [3]. Most organisms use DNA for their long-term information storage, but some viruses (e.g., retroviruses) have RNA as their genetic material. The biological information contained in an organism is encoded in its DNA or RNA sequence. RNA is also used for information transport (e.g., mRNA) and enzymatic functions (e.g., ribosomal RNA) in organisms that use DNA for the genetic code itself [23]. Transfer RNA (tRNA) molecules are used to add specific amino acids during the process of protein translation. Prokaryotic genetic material is organized in a simple circular DNA molecule (the bacterial chromosome) in the nucleoid region of the cytoplasm. Eukaryotic genetic material is divided into different, linear molecules called chromosomes inside a discrete nucleus, usually with additional genetic material in some organelles like mitochondria and chloroplasts [24]. A human cell has genetic material in the nucleus (the nuclear genome) and in the mitochondria (the mitochondrial genome). In humans the nuclear genome is divided into 23 pairs of linear DNA molecules called chromosomes [25]. The mitochondrial genome is a circular DNA molecule distinct from the nuclear DNA. Mitochondrial DNA codes 13 proteins involved in mitochondrial energy production as well as specific tRNAs, although mitochondrial DNA is very small compared to nuclear chromosomes. Foreign genetic material (most commonly DNA) can also be artificially introduced into the cell by a process called transfection. Certain viruses also insert their genetic material into the genome.

2.1.3.4 Organelles

The human body contains many different organs, such as the heart, lung, and kidneys, with each organ performing a different function. Cells also have a set of "little organs," called organelles that are adapted and/or specialized for carrying out one or more vital functions [26]. There are several types of organelles within an animal cell. Some (such as the nucleus and Golgi apparatus) are typically solitary, while others (such as mitochondria, peroxisomes, and lysosomes) can be numerous (hundreds to thousands). The cytosol is the gelatinous fluid that fills the cell and surrounds the organelles.

Mitochondria and Chloroplasts

Mitochondria are self-replicating organelles that occur in various numbers, shapes, and sizes in the cytoplasm of all eukaryotic cells. Mitochondria play a critical role in generating energy in the eukaryotic cell. Mitochondria generate the cell's energy by the process of oxidative phosphorylation, utilizing oxygen to release energy stored in cellular nutrients (typically pertaining to glucose) to generate ATP. Mitochondria multiply by splitting in two. Organelles that are modified chloroplasts are broadly called plastids, and are involved in energy storage through the process of photosynthesis, which utilizes solar energy to generate carbohydrates and oxygen from carbon dioxide and water [1]. Mitochondria and chloroplasts each contain their own genome, which is separate and distinct from the nuclear genome of a cell. Both of these organelles contain this DNA in circular plasmids, much like prokaryotic cells, strongly supporting the evolutionary theory of endosymbiosis; since these organelles contain their own genomes and have other similarities to prokaryotes, they are thought to have developed through a symbiotic relationship after being engulfed by a primitive cell [2, 3].

Ribosomes

The ribosome is a large complex of RNA and protein molecules [3]. They each consist of two subunits, and act as an assembly line where mRNA from the nucleus is used to synthesize proteins from amino acids. Ribosomes can be found either floating freely or bound to a membrane (i.e., in eukaryotes, which can be found in the rough endoplasmatic reticulum, and in prokaryotes, which can be found in the cell membrane) [27].

2.1.3.5 Cell Nucleus

The cell nucleus is the most conspicuous organelle found in a eukaryotic cell. It houses the cell's chromosomes; almost all DNA replication and RNA synthesis (transcription) occur in the cell nucleus [1]. The nucleus is spherical in shape and separated from the cytoplasm by a double membrane called the nuclear envelope. The nuclear envelope isolates and protects a cell's DNA from various molecules that could accidentally damage its structure or interfere with its processing. During processing, DNA is transcribed, or copied into a special RNA, called mRNA. This mRNA is then transported out of the nucleus, where it is translated into a specific protein molecule. The nucleolus is a specialized region within the nucleus where ribosome subunits are assembled. In prokaryotes, DNA processing takes place in the cytoplasm [23].

2.2 Electromechanics of Particles

As a result of the particles' electrical and magnetic properties, they experience force and torques when subjected to electrical and magnetic fields. Moreover, mutual interactions between the particles and fields are obtained when the particles are electrically charged, polarized, or magnetized [28–33]. The particle-field interactions are referred to as the electromecanics of particles.

2.2.1 Single-Layer Model

For prokaryotic cells where there is no nucleus such as bacteria (see Section 2.1.1), a single-layer model can be used. Figure 2.3 shows a single-layer model [33]. It shows two different regions; they are:

1. The cell membrane, which consists of a selective bilayer of lipid protein molecules;
2. The cell interior (cytoplasm).

From the electrical point of view, the membrane is typically characterized by effective capacitance C_{cm} and conductance g_{cm}. The cell interior (cytoplasm) is simplified as a homogenous model with permittivity C_c and ohmic conductivity σ_c.

2.2.2 Double-Layer Model

In a eukaryotic cell where there is a nucleus such as in animal or plant, a double-layer model can be adapted [33]. This model is shown in Figure 2.4. This figure shows four different regions:

1. The wall of the cell (cell membrane).
2. The cell interior (cytoplasm).
3. The nuclear membrane, which consists of a selectively bilayer of lipid protein molecules. The nuclear membrane serves as a two-way conduit for life-sustaining nutrients and substances required by the cell. It also passes the waste materials outside of the cell.
4. Nucleoplasm, an aqueous fluid that contains the nucleus and other structures.

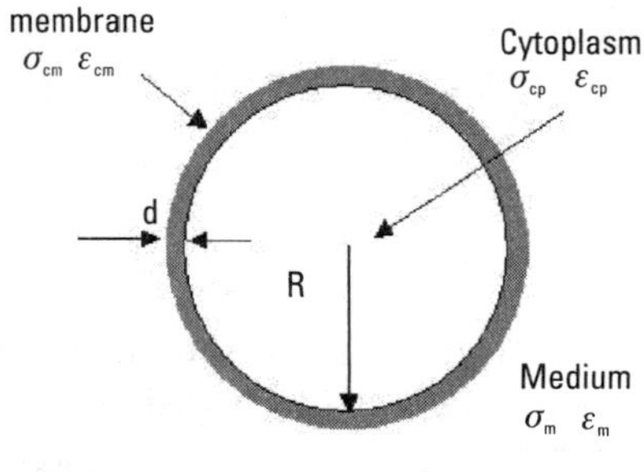

Figure 2.3 Single-layer model.

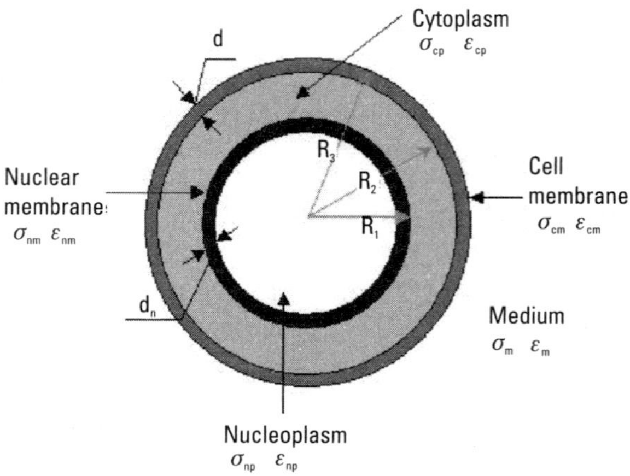

Figure 2.4 Double-layer model.

The wall helps provide rigidity to the cell. For example, plant tissue consists of a large numbers of these cells standing together by the wall material, which gives structural rigidity to leaves and other plant parts. Another example is the yeast, which is a single-cell organism, which has a roughly spherical shape. In yeasts, the cell wall provides mechanical protection to the membrane. From the electrical point of view, the wall of the cell is usually modeled as a homogenous spherical concentric shell of finite thickness with bulk permittivity ϵ_w and ohmic conductivity σ_w. The membrane is typically characterized by effective capacitance C_m and conductance g_m. The cell interior (cytoplasm) is simplified as a homogenous model with permittivity ϵ_{cm} and ohmic conductivity σ_{cm}. Also, the nucleoplasm is simplified as a homogenous model with permittivity ϵ_{np} and ohmic conductivity σ_{np}.

2.3 Electrogenic Cells

Cells that have the ability to generate electrical signals are called electrogenic cells. In the human body, there are many types of cells that can go in a transient depolarization and repolarization (i.e., generate electric signals) such as brain cells, neurons, and hart cells [34].

2.3.1 Neurons

Neurons (nerve cells) are biological cells that specialize in transmitting and processing information. The elementary neural signals are action potentials, which are transit charges of voltage drop across the cell membrane with a typical shape. The human brain contains a large number of neurons (in the tens of billions) and about 1,000 times as many synaptic connections [35–37]. A high degree of connectivity and the coordinated electrical activity of neural networks are believed to be the reason that the brain is capable of highly sophisticated information processing.

The neuron has four distinct regions:

1. The *cell body*, which has a diameter of 10 to 100 μm. It contains the nucleus and produces most neuronal proteins.
2. The *dendrites*, which conduct the electrochemical stimulation received from other neural cells to the cell body of the neuron from which the dendrites project. Neurons have multiple dendrites that extend outward from the cell body and are specialized to receive either chemical or electrical signals from other neurons via synapses. Dendrites convert theses signals into electrical impulses and transmit them to the cell body.
3. The *axon*, which is specialized for the conduction of electrical pulses, termed action potentials, away from the cell body towards the axon terminals.
4. The *axon terminals*, which are the last part in the axon and make contact with other cells.

Figure 2.5 shows the neuron's structure. Neurons communicate with other neurons through specialized contacts called synapses. Synapses can be either chemical or electrical, and have two main functions in the transmission of impulses from one cell to the other. These functions are: signal amplification, which can be observed at nerve-muscle synapses where synapses amplify the signal received from the nerve and stimulate contraction; and signal computation, which is common at a synapse's interneuron. For example, a single neuron can be affected simultaneously by signals received from different excitatory synapses. The neuron averages these signals and determines if an action potential generation is needed or not. The synapse plays the main role for computation; this is the basis of the ability of neural networks to analyze and perform computation.

2.3.2 Gated Ion Channels

Ion channels are embedded transmembrane proteins that help establish and control the small voltage gradient across the plasma membrane of all living cells by allowing the flow of ions down their electrochemical gradient [38]. They are present in the membranes that surround all biological cells. In other words, ion channels allow the formation of a concentration gradient of ions between the extracellular and intracellular fluids surrounding the membrane. This concen-

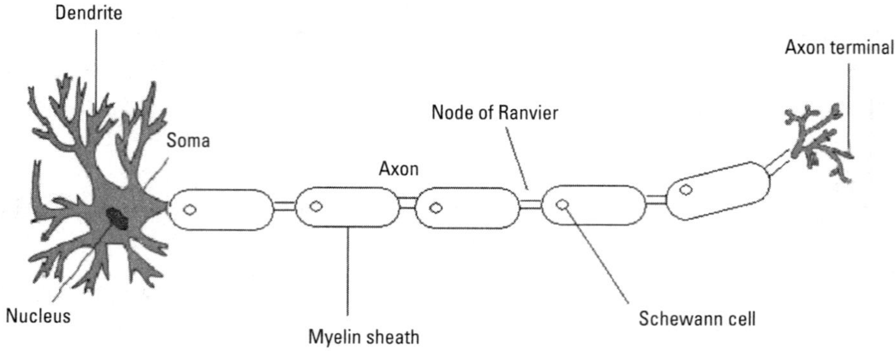

Figure 2.5 Neuron structure.

tration gradient leads to an electrical action potential across the cell membrane (typically between −30 to −70 mV). If the ion channels can open or close then they are called gated ion channels. There are three types of gated ion channels: ligand-gated channels, mechanically gated channels, and voltage-gated channels [2].

2.3.2.1 Hodgkin-Huxley Model

The best-studied model of the mechanisms in an excitable membrane is the Hodgkin-Huxley model, which is a scientific model that describes how action potentials in neurons are initiated and propagated [39]. The Hodgkin-Huxley model is a set of nonlinear ordinary differential equations that approximates the electrical characteristics of excitable cells such as neurons and cardiac cells. In this model, the ion currents are flowing through tansmembrane channel proteins through which ions can diffuse down their concentration gradients. These tansmembrane channels have gates that are controlled by voltage-sensitive gating charges.

The components of a typical Hodgkin-Huxley model are shown in Figure 2.6. Each component of an excitable cell has a biophysical analogy [39]. The lipid bilayer is represented as a capacitance (C_m). Voltage-gated ion channels are represented by nonlinear electrical conductances (g_n), where n is the specific ion channel, meaning that the conductance is voltage- and time-dependent. This was later shown to be mediated by voltage-gated cation channel proteins, each of which has an open probability that is voltage-dependent. Leak channels are represented by linear conductances (g_L). The electrochemical gradients driving the flow of ions are represented by batteries (E_n and E_L), the values of which are determined from the Nernst potential of the ionic species of interest. Finally, ion pumps are represented by current sources (I_p).

To simplify the complexity of the mathematics used in Hodgkin-Huxley model, a Fitzhugh-Nagumo (FHN) model is developed. The FHN model has

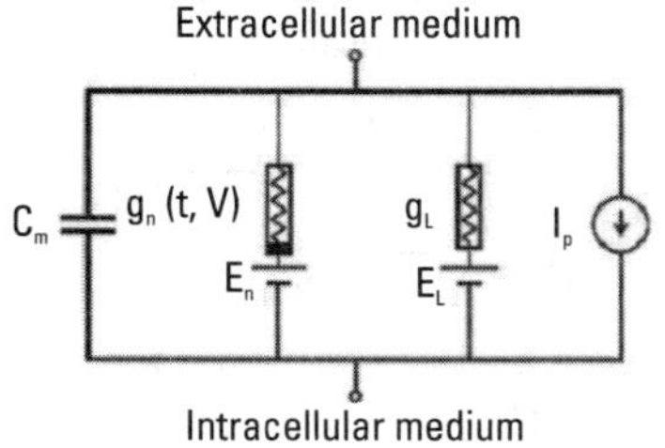

Figure 2.6 Hodgkin-Huxley model [39].

only two independent variables, but exhibits a similar stability behavior to the full Hodgkin-Huxley equations [40].

A hybrid of the Hodgkin-Huxley and Fitzhugh-Nagumo models was developed by Morris and Lecar in 1981 [41]. The Morris-Lecar model replaces the voltage-gated sodium current of the Hodgkin-Huxley model with a voltage-dependent calcium current.

2.3.3 Action Potential

The activity of voltage-gated ion channels embedded in the membrane generates an action potential (nerve impulse), which is a transient alteration of the transmembrane voltage (membrane potential) across an excitable membrane. Action potentials play multiple roles in several types of excitable cells such as neurons, myocytes, and electrolytes. The action potentials are pulse-like waves of voltage that travel along axons of neurons [42]; Figure 2.7 shows the action potential.

A typical action potential is initiated at the axon hillock when the membrane is depolarized sufficiently (i.e., when its voltage is increased sufficiently) [43–45]. The procedure of an action potential can be summarized as follows:

1. The membrane potential is increased.
2. Sodium ion channels open, allowing the entry of sodium ions into the cell.
3. Potassium ion channels open, which permits the exit of potassium ions from the cell.
4. The inward flow of sodium ions increases the concentration of positively charged cations in the cell and causes depolarization, where the potential of the cell is higher than the cell's resting potential.
5. The sodium channels close at the peak of the action potential, while potassium continues to leave the cell. The efflux of potassium ions decreases the membrane potential or hyperpolarizes the cell.

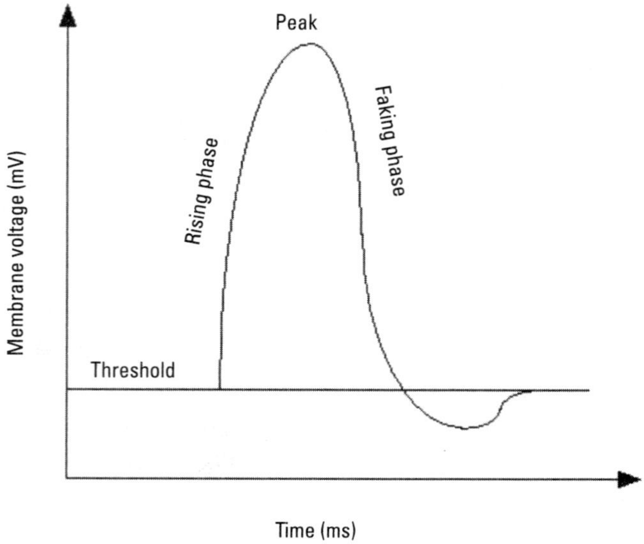

Figure 2.7 Schematic action potential signal.

6. For small voltage increases from rest, the potassium current exceeds the sodium current and the voltage returns to its normal resting value, typically −70 mV [43]. However, if the voltage increases past a critical threshold, typically 15-mV higher than the resting value, the sodium current dominates. This results in a runaway condition whereby the positive feedback from the sodium current activates even more sodium channels. Thus, the cell produces an action potential [45].

Currents produced by the opening of voltage-gated channels in the course of an action potential are typically significantly larger than the initial stimulating current. Thus the properties of the excitable membrane are controlling the amplitude, duration, and shape of the action potential. These parameters are not determined by the amplitude or duration of the stimulus. This all-or-nothing property of action potential sets it apart from graded potentials such as receptor potentials, electrotonic potentials, and synaptic potentials, which scale with the magnitude of the stimulus. A variety of action potential types exist in many cell types and cell compartments as determined by the types of voltage-gated channels, leak channels, channel distributions, ionic concentrations, membrane capacitance, temperature, and other factors [43–45].

The principal ions involved in an action potential are sodium and potassium cations; sodium ions enter the cell, and potassium ions leave, restoring equilibrium [43]. Relatively few ions need to cross the membrane for the membrane voltage to change drastically. The ions exchanged during an action potential, therefore, make a negligible change in the interior and exterior ionic concentrations. The few ions that do cross are pumped out again by the continual action of the sodium–potassium pump, which, with other ion transporters, maintains the normal ratio of ion concentrations across the membrane. In a domino-like propagation, action potentials are generated locally on patches of excitable membrane and the resulting currents can trigger action potentials on neighboring stretches of membrane. Action potentials are generated along excitable stretches of membrane and propagate without decay [45], contrarily to the passive spread of electric potentials (electrotonic potential).

References

[1] Campbell, N. A., B. Williamson, and R. J. Heyden, *Biology: Exploring Life*, Upper Saddle River, NJ: Prentice-Hall, 2006.

[2] Maton, A., et al., *Cells Building Blocks of Life*, Upper Saddle River, NJ: Prentice-Hall, 1997.

[3] Alberts, B., *Molecular Biology of the Cell*, 4th ed., New York: Garland Science, 2002.

[4] Ehrlicher, A. R., "The Forces Behind Cell Movement," *International Journal of Biological Science*, Vol. 3, 2007, pp. 303–317.

[5] Orgel, L., "The Origin of Life—A Review of Facts and Speculations," *Trends Biochem. Sci.*, Vol. 23, 1998, pp. 491–495.

[6] Griffiths, G., "Cell Evolution and the Problem of Membrane Topology," *Nature Reviews, Molecular Cell Biology*, Vol. 8, 2007, pp. 1018–1024.

[7] Sterrer, W., "On the Origin of Sex as Vaccination," *Journal of Theoretical Biology*, Vol. 216, 2002, pp. 387–396.

[8] Kaiser, D., "Coupling Cell Movement to Multicellular Development in Myxobacteria," *Nat. Rev. Microbiol*, Vol. 1, 2003, pp. 45–54.

[9] Sapp, J., "The Prokaryote-Eukaryote Dichotomy: Meanings and Mythology," *Microbiol. Mol. Biol. Rev.*, Vol. 69, 2005, pp. 292–305.

[10] Thanbichler, M., S. Wang, and L. Shapiro, "The Bacterial Nucleoid: A Highly Organized and Dynamic Structure," *J. Cell Biochem.*, Vol. 96, 2005, pp. 506–521.

[11] Harold, F., "Conservation and Transformation of Energy by Bacterial Membranes," *Bacteriol. Rev.*, Vol. 36, 1972, pp. 172–230.

[12] Shih, Y., and L. Rothfield, "The Bacterial Cytoskeleton," *Microbiol. Mol. Biol. Rev.*, Vol. 70, 2006, pp. 729–754.

[13] Michie, K., and J. Löwe, "Dynamic Filaments of the Bacterial Cytoskeleton," *Annu. Rev. Biochem.*, Vol. 75, 2006, pp. 467–492.

[14] Fuerst, J., "Intracellular Compartmentation in Planctomycetes," *Annu. Rev. Microbiol.*, Vol. 59, 2005, pp. 299–328.

[15] Ciccarelli, F., et al., "Toward Automatic Reconstruction of a Highly Resolved Tree of Life," *Science*, Vol. 311, 2006, pp. 1283–1287.

[16] Knoll, A. H., et al., "Eukaryotic Organisms in Proterozoic Oceans," *Philosophical Transactions of the Royal Society of London, Part B*, Vol. 361, No. 1470, 2006, pp. 1023–1038.

[17] Brocks, J., et al., "Archean Molecular Fossils and the Early Rise of Eukaryotes," *Science*, Vol. 285, 1999, pp. 1033–1036.

[18] Ward, P., "Mass Extinctions: The Microbes Strike Back," *New Scientist*, 2008, pp. 40–43.

[19] Tovar, J., A. Fischer, and C. Clark, "The Mitosome, a Novel Organelle Related to Mitochondria in the Amitochondrial Parasite Entamoeba Histolytica," *Mol. Microbiol.*, Vol. 32, 1999, pp. 1013–1021.

[20] Frixione, E., "Recurring Views on the Structure and Function of the Cytoskeleton: A 300-Year Epic," *Cell Motility and the Cytoskeleton*, Vol. 46, No. 2, 2002, pp. 73–94.

[21] Minton, A., "Confinement as a Determinant of Macromolecular Structure and Reactivity," *Journal of Biophys.*, Vol. 63, No. 4, 1992, pp. 1090–1100.

[22] Doherty, G., and H. McMahon, "Mediation, Modulation and Consequences of Membrane-Cytoskeleton Interactions," *Annual Review of Biophysics*, Vol. 37, 2008, pp. 65–95.

[23] Ghosh, A., and M. Bansal, "A Glossary of DNA Structures from A to Z," *Acta. Crystallogr. D Biol. Crystallogr.*, Vol. 59, 2003, pp. 620–626.

[24] Wing, R., et al., "Crystal Structure Analysis of a Complete Turn of B-DNA," *Nature*, Vol. 287, No. 5784, 1980, pp. 755–758.

[25] Pabo, C., and R. Sauer, "Protein-DNA Recognition," *Annu. Rev. Biochem.*, Vol. 53, 1984, pp. 293–321.

[26] Frey-Wyssling, A., "Concerning the Concept 'Organelle,'" *Experientia*, Vol. 34, No. 547, 1978.

[27] Recht, M., S. Douthwaite, and J. Puglisi, "Basis for Bacterial Specificity of Action of Aminoglycoside Antibiotics," *EMBO Journal*, Vol. 18, No. 11, 1999, pp. 3133–3138.

[28] Schwan, H. P., "Dielectrophoresis and Rotation of Cell," in E. Neumann, A. E. Sowers, and C. A. Jordan, (eds.), *Electroporation and Electrofusion in Cell Biology*, New York: Plenum Press, 1989.

[29] Zimmermann, U., and G. A. Neil, *Electromanipulation of Cells*, Boca Raton, FL: CRC Press, 1996.

[30] Schoenbach, K. H., et al., "Bioelectrics—New Applications for Pulsed Power Technology," *IEEE Trans. on Plasma Science*, Vol. 30, No. 1, 2002.

[31] Dev, S. B., et al., "Medical Applications of Electroporation," *IEEE Trans. on Plasma Science*, Vol. 28, No. 1, 2000, pp. 206–223.

[32] Jaroszeski, M., R. Heller, and R. Gilbert, (eds.), *Electrochemotherapy, Electrogenetherapy, and Transdermal Drug Delivery: Electrically Mediated Delivery of Molecules to Cells, Methods in Molecular Medicine*, Vol. 37, Totowa, NJ: Humana Press, 2000.

[33] Jones, T. B., *Electromechanics of Particles*, Cambridge, U.K.: Cambridge University Press, 1995.

[34] Lee, H., D. Ham, and R. W. Westervelt, *CMOS Biotechnology*, New York: Springer, 2007.

[35] Johnston, D., and S. Wu, *Foundations of Cellular Neurophysiology*, Cambridge, MA: MIT Press, 1995.

[36] Lodish, H. F., *Molecular Cell Biology*, 3rd ed., New York: Scientific American Books, 1995.

[37] Kandel, E. R., J. H. Schwartz, and T. M. Jessel, *Principles of Neural Science*, 4th ed., New York: McGraw-Hill, 2000.

[38] Hille, B., *Ion Channels of Excitable Membranes*, 3rd ed., Sunderland, MA: Sinauer Associates, 2001.

[39] Hodgkin, A., and A. Huxley, "A Quantitative Description of Membrane Current and Its Application to Conduction and Excitation in Nerve," *J. Physiol.*, 1952, pp. 500–544.

[40] Rinzel, J., and G. Ermentrout, "Analysis of Neural Excitability and Oscillations," in C. Koch and I. Segev, (eds.), *Methods in Neuronal Modeling: From Synapses to Networks*, Cambridge, MA: MIT Press, 1989, pp. 135–169.

[41] Morris, C., and H. Lecar, "Voltage Oscillations in the Barnacle Giant Muscle Fiber," *Biophysical Journal*, Vol. 35, 1981, pp. 193–213.

[42] Mummert, H., and D. Gradmann, "Action Potentials in Acetabularia: Measurement and Simulation of Voltage-Gated Fluxes," *Journal of Membrane Biology*, Vol. 124, 1991, pp. 265–273.

[43] Doyle, D., et al., "The Structure of the Potassium Channel, Molecular Basis of K+ Conduction and Selectivity," *Science*, Vol. 280, 1998, pp. 69–77.

[44] Zhou, Y., et al., "Chemistry of Ion Coordination and Hydration Revealed by a K+-Fab Complex at 2.0 A Resolution," *Nature*, Vol. 414, 2001, pp. 43–48.

[45] Jiang, Y., et al., "X-Ray Structure of a Voltage-Dependent K+ Channel," *Nature*, Vol. 423, 2003, pp. 33–41.

3

Cell Manipulator Fields

Particles have electrical and magnetic properties associated with their material and shapes. Consequently, they experience forces and torques when subjected to electrical and/or magnetic fields. Moreover, particles will exhibit mutual interactions through their own electric charge, polarization, or magnetization.

Due to the action of electric and magnetic fields, forces and torques are exerted on small particles (less than a millimeter in diameter) [1]. The electric or magnetic field can be obtained either by external means (i.e., electrodes or magnetic poles) or by other nearby charges or magnetized particles.

In this chapter, different types of electric and magnetic fields will be presented and discussed.

3.1 Electric Field

3.1.1 Uniform Electric Field (Electrophoresis)

Electrophoresis is the motion of dispersed particles relative to a fluid under the influence of a spatially uniform electric field [2]. In other words, electrophoresis (EP) is the method of moving charged particles through a medium by using an electric field induced by flat electrodes [3]. The features of the EP phenomenon can be summarized as follows:

- Particles experience EP force only when the electric field is uniform.

- The EP force depends on the polarity of the applied electric field and is observed with DC excitation only.
- There are two kinds of EP forces:
    - *Positive EP for positive charged particles.* Particles move in the same direction as the electric field.
    - *Negative EP for negative charged particles.* Particles move in the opposite direction against the electric field.

3.1.2 Nonuniform Electric Field (Dielectrophoresis)

Dielectrophoresis (DEP) [1, 4] is defined as the motion of uncharged (neutral) particles caused by the polarization effect in a nonuniform electric field. The features of the DEP phenomenon can be summarized as follows:

- Particles experience DEP force only when the electric field is nonuniform.
- The DEP force does not depend on the polarity of the applied electric field and is observed with AC as well as DC excitation. There are two kinds of DEP forces:
    - Positive DEP for $\varepsilon_m < \varepsilon_p$, where ε_m is the permittivity of the suspended medium and ε_p is the permittivity of the particles. In this case, particles are attracted to regions of the stronger electric field.
    - Negative DEP for $\varepsilon_m > \varepsilon_p$. In this case, particles are repelled from regions of the stronger electric field.
- DEP is most readily observed for particles with diameters ranging from approximately 1–1,000 μm.

DEP should be contrasted with electrophoresis, where one manipulates charged particles in a dissipative medium with electric fields [5], as there are several important differences. First, DEP does not require the particle to be charged in order to manipulate it. The particle must only differ electrically from the medium that it is in. Second, DEP works with AC fields, where no net electrophoretic movement occurs in such a field. Thus, with DEP one can avoid problems such as electrode polarization effects and electrolysis at electrodes. Even more importantly, the use of AC fields reduces membrane charging of biological cells [1]. Membrane charging is due to the potential developed across cell membranes in electric fields. This potential, which can impact cell physiology, can be diminished by the application of high-frequency fields. Finally, DEP forces increase with the gradient of the square of the electric field (i.e., $\nabla\left|\vec{E}\right|^2$, described later), whereas electrophoretic forces increase linearly with the electric field.

To illustrate these differences, Figure 3.1 shows a uniform electric field applied to two bodies; one of them is neutral and the second is charged. In a uniform electric field, a charged body is pulled towards the electrode carrying the charge opposite to that on the particle. In the same field, a neutral body will merely be polarized. The result may produce a torque, but not a translational force; that is, the body will not move toward either electrode. In Figure 3.2, where a nonuniform electric field is shown, we can observe different behavior of the charged and neutral bodies. The charged body is still attracted toward the electrode of opposite polarity. The neutral body, in this case, will find a translational force upon it. This occurs as a result of the polarization, which causes a physical displacement of charges in the dielectric [4], which in turn happens under the influence of the nonuniform electric field. The polarization has the effect of putting a negative charge upon the side nearer the positive electrode, and a positive one on the side nearer the negative electrode. Because the particle is neutral, the two charges on the body are opposite but equal. Figure 3.2 shows the reaction of both neutral and charged particles under the effect of a nonuniform electric field. To characterize the DEP force, the Dipole theory has been used [1].

3.1.2.1 Dielectrophoreses (DEP) Force

In this section, an estimated expression of the net force upon a small physical dipole will be obtained [1, 4]. The dipole consists of equal and opposite charges $+q$ and $-q$ located a vector distance $\vec{d}$ apart, under the effect of a nonuniform electric field $\vec{E}(\vec{r})$, as shown in Figure 3.3. The dipole moment $\vec{p}$ is defined as the result of multiplication of the charge q and the distance $\vec{d}$ (i.e., $\vec{p}-q\vec{d}$). It is assumed that $|\vec{d}|$ is very small compared to the characteristic dimensions of the electric field nonuniformity, and the electric field $\vec{E}(\vec{r})$ includes no contributions due to the dipole itself.

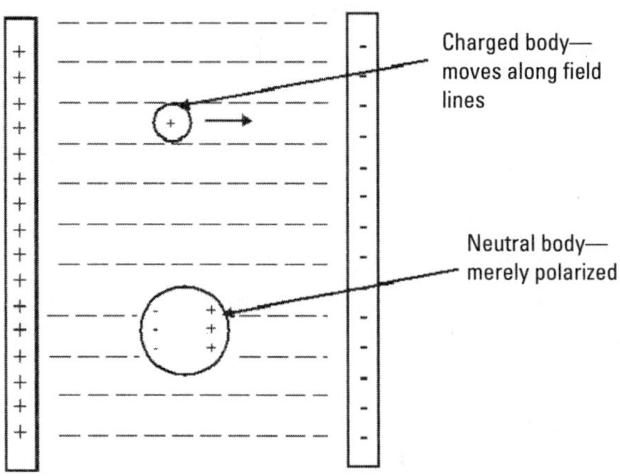

Figure 3.1 Uniform electric field applied to neutral and charged bodies.

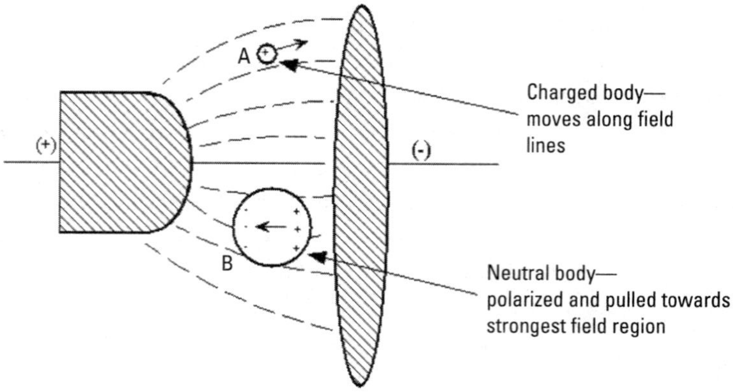

Figure 3.2 Nonuniform electric field applied to neutral and charged bodies.

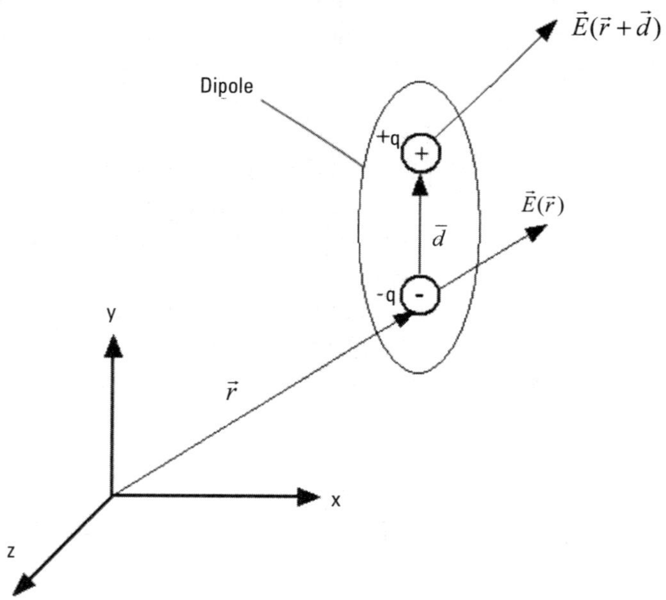

Figure 3.3 Representation of an elementary dipole in a nonuniform electric field.

Because the applied electric field $\vec{E}(\vec{r})$ is nonuniform, then, in general, the two charges ($+q$ and $-q$) will experience different values of the vector field, $\vec{E}(\vec{r})$ and $\vec{E}(\vec{r}+\vec{d})$, respectively, and the dipole will experience a net force. The charge $+q$ experiences

$$\vec{F}_{+} = q\vec{E}\left(\vec{r}+\vec{d}\right) \tag{3.1}$$

while the charge $-q$ experiences

$$\vec{F}_{-} = -q\vec{E}(\vec{r}) \tag{3.2}$$

The net force on the dipole is:

$$\vec{F} = q\vec{E}\left(\vec{r}+\vec{d}\right) - q\vec{E}(\vec{r}) \tag{3.3}$$

Using a Taylor series expansion,

$$\vec{E}\left(\vec{r}+\vec{d}\right) = \vec{E}(\vec{r}) + \vec{d}.\nabla\vec{E}(\vec{r}) + \text{ higher-order terms} \tag{3.4}$$

where $\nabla\vec{E}(\vec{r})$ is the gradient of $\vec{E}(\vec{r})$.

Taking into consideration that $\left|\vec{d}\right|$ is very small compared to the characteristic dimension of the electric field nonuniformity, the higher additional terms, of order d^2, d^3, and so forth in (3.4) have been neglected. Substituting (3.4) into (3.3), we can simplify (3.3) as given here:

$$\vec{F} = q\vec{d}.\nabla\vec{E} \tag{3.5}$$

As $\vec{p} - q\vec{d}$ (the dipole moment), the force on an infinitesimal dipole results in [1]:

$$\vec{F}_{dipole} = \vec{p}.\nabla\vec{E} \tag{3.6}$$

Similarly, if we have a neutral particle suspended in some dielectric fluid and polarized by a nonuniform electric field $\vec{E}$, and due to polarization, $\vec{E}$ will induce a moment in the particle and a net electric force will be given as:

$$\vec{F} = \left(\vec{p}_{eff}.\nabla\right)\vec{E} \tag{3.7}$$

where $\vec{p}_{eff}$ is the effective dipole moment, and is defined as the moment of an equivalent, free charge, point dipole that when immersed in the same dielectric fluid and positioned at the same location as the center of the original particle, produces the same dipolar electrostatic potential.

In the case of a sphere in a medium, both of which are lossless dielectrics (i.e., both materials have finite dielectric constant and zero conductivity (ideal case), the induced effective dipole moment $\vec{p}_{eff}$ is given by [1].

$$\vec{p}_{eff} = 4\pi a^3 \varepsilon_m \left(\frac{\varepsilon_p - \varepsilon_m}{\varepsilon_p + 2\varepsilon_m} \right) \vec{E} \tag{3.8}$$

where a is the radius of the sphere, ε_m is the permittivity of the medium, ε_p is the permittivity of the sphere, and $\vec{E}$ is the strength of the applied electric field. Substituting (3.8) into (3.7), the net force on the sphere is:

$$\vec{F} = 2\pi a^3 \varepsilon_m \left(\frac{\varepsilon_p - \varepsilon_m}{\varepsilon_p + 2\varepsilon_m} \right) \nabla \left| \vec{E} \right|^2 \tag{3.9}$$

The term $\left(\frac{\varepsilon_p - \varepsilon_m}{\varepsilon_p + 2\varepsilon_m} \right)$ is known as the Clausius-Mossotti factor (K) [1].

In the nonideal case, where ohmic losses are taken into consideration, a homogenous spherical particle with radius a and complex permittivity ω_{cp}, when immersed in a medium of complex permittivity ε_m^* and exposed to a spatially nonuniform AC electric field, the time average DEP force $\left\langle \vec{F}_{DEP} \right\rangle$ is defined as follows:

$$\left\langle \vec{F}_{DEP} \right\rangle = 2\pi a^3 \varepsilon_m \operatorname{Re}[\underline{K}] \nabla E_{rms}^2 \tag{3.10}$$

where

$$\underline{K} = \left(\frac{\varepsilon_p^* - \varepsilon_m^*}{\varepsilon_p^* + 2\varepsilon_m^*} \right) \tag{3.11}$$

is the complex Clausius-Mossotti factor, and ε_p^* and ε_m^* are defined by (3.12) and (3.13), respectively:

$$\varepsilon_p^* = \varepsilon_p - j\frac{\sigma_p}{\omega} \tag{3.12}$$

$$\varepsilon_m^* = \varepsilon_m - j\frac{\sigma_m}{\omega} \tag{3.13}$$

where σ_p and σ_m are the conductivity of the particle and the medium, respectively, and ω is the angular frequency of the applied electric field. From (3.10), we can observe the following:

- The DEP force is linearly related to the volume of the immersed particle (i.e., $F_{DEP} \alpha a^3$).
- F_{DEP} is related directly to the permittivity of the suspended medium (i.e., $F_{DEP} \alpha \varepsilon_m$).
- The DEP force depends upon the magnitude and sign of the Clausius-Mossotti factor.
- The DEP force vector is directed along the gradient of the electric field intensity ∇E_{rms}^2, and the time-averaged DEP force is independent of the field polarity, as we are taking the rms value of the intensity of the electric field.

3.1.2.2 Positive and Negative DEP Forces

In accordance with (3.10), the real part of the Clausius-Mossotti factor ($\text{Re}[\underline{K}]$) determines the direction of the DEP force. When $\text{Re}[\underline{K}]$ is positive (i.e., $\varepsilon_m < \varepsilon_p$), particles are attracted to the electric field intensity maxima and repelled from the electric field intensity minima. This force is known as *positive dielectrophoresis. Negative dielectrophoresis* occurs when $\text{Re}[\underline{K}] < 0$ (i.e., $\varepsilon_m > \varepsilon_p$), where particles are attracted to the electric field intensity minima and repelled from the electric field intensity maxima.

The most interesting dielectrophoretic behavior is exhibited by particles exhibiting frequency-dependent dispersion due to dielectric or conductive loss. Substituting (3.12) and (3.13) into (3.11), we obtain the frequency dependence of $\text{Re}[\underline{K}]$ [1]:

$$\text{Re}[\underline{K}] = \frac{\varepsilon_p - \varepsilon_m}{\varepsilon_p + 2\varepsilon_m} + \frac{3\left(\varepsilon_m \sigma_p - \varepsilon_p \sigma_m\right)}{\tau_{MW}\left(\sigma_p + 2\sigma_m\right)^2\left(1 + \omega^2 \tau_{MW}^2\right)} \tag{3.14}$$

where $\tau_{MW} = \dfrac{\varepsilon_p + \varepsilon_m}{\sigma_p + 2\sigma_m}$ is the Maxwell-Wagner charge relaxation time.

The high and low frequency limits for $\text{Re}[\underline{K}]$ will be identified as:

$$\text{Re}[\underline{K}] = \frac{\sigma_p - \sigma_m}{\sigma_p + 2\sigma_m}, \quad \text{for } \omega \to 0 \tag{3.15}$$

$$\text{Re}[\underline{K}] = \frac{\varepsilon_p - \varepsilon_m}{\varepsilon_p + 2\varepsilon_m}, \quad \text{for } \omega >> 0 \tag{3.16}$$

We can observe from (3.15) and (3.16) that the conductivity dominates the low frequency DEP behavior, while the permittivity dominates the high frequency behavior. Figures 2.4 and 2.5 show a schematic diagram representation of Re[$\underline{K}$] versus radian frequency ω for two different cases. In the first case (Figure 3.4), where $\sigma_p < \sigma_m$ and $\varepsilon_p > \varepsilon_m$, Re[$\underline{K}$] is negative at low frequencies and positive at high frequencies. The reverse case is shown in Figure 3.5, where $\sigma_p > \sigma_m$ and $\varepsilon_p < \varepsilon_m$, so that Re[$\underline{K}$] becomes positive at low frequencies and negative at high frequencies. Thus, the frequency dependence of Re[$\underline{K}$] translates to a frequency-dependent, time-averaged DEP force $\langle \vec{F}_{DEP} \rangle$ on homogenous particles (3.10). In other words, both the magnitude and sign of $\langle \vec{F}_{DEP} \rangle$ are functions of the electric field frequency ω.

For example, in Figure 3.4, where $\sigma_p < \sigma_m$ and $\varepsilon_p > \varepsilon_m$ the particle will be attracted to the electric field intensity minima at low frequencies and the maxima at high frequencies. The crossover frequency ω_c (Figures 3.4 and 3.5), at which no DEP force acts on a particle (i.e., Re[$\underline{K}$] = 0), and the DEP force changes sign can be calculated as [1].

$$\omega = \sqrt{\frac{(\sigma_m - \sigma_p)(\sigma_p + 2\sigma_m)}{(\varepsilon_p - \varepsilon_m)(\varepsilon_p + 2\varepsilon_m)}} \tag{3.17}$$

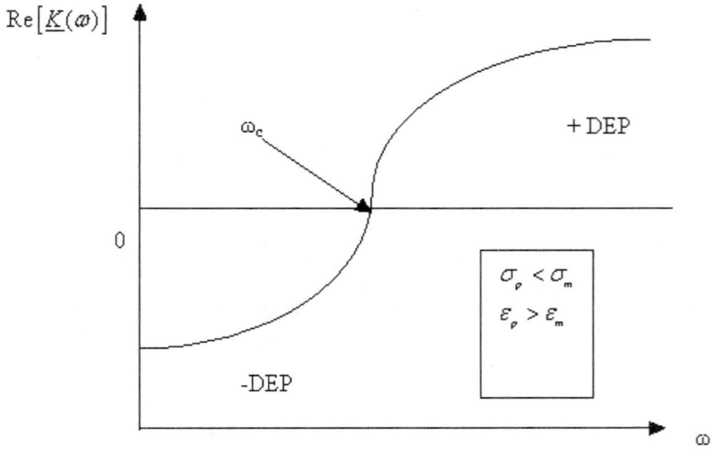

Figure 3.4 Dielectrophoretic spectra for $\sigma_p < \sigma_m$ and $\varepsilon_p > \varepsilon_m$.

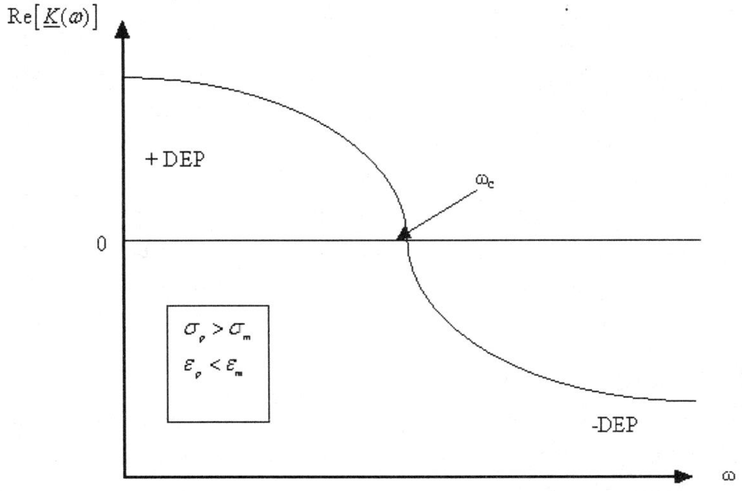

Figure 3.5 Dielectrophoretic spectra for $\sigma_p > \sigma_m$ and $\varepsilon_p < \varepsilon_m$.

It can be deduced that ω_c is defined only when $\frac{\sigma_m - \sigma_p}{\varepsilon_p - \varepsilon_m} > 0$ (i.e., the DEP spectrum crosses the *x*-coordinate).

3.1.2.3 Dielectrophrtic (DEP) Levitation

Both positive DEP and negative DEP can be used to precisely displace cells in a microchamber formed between two facing glass chips with elongated electrodes [6]. However, this procedure has the disadvantage that cells tend to contact the device surfaces and stick to the surfaces. A solution to this problem is to levitate cells while manipulating them using dielectrophoretic levitation (DEP levitation). The DEP levitation technique is based on the balance of the gravitational force and the DEP force where a particle becomes stably suspended in a fluid of known properties [7, 8]. To levitate a particle exhibiting negative DEP (passive levitation) ($\mathrm{Re}[\underline{K}] < 0$), requires a local minimum in the electric field intensity. Such minima are observed in divergence and curl-free electrostatic fields [1]. Thus passive levitation of particles exhibiting negative DEP is readily achieved with cusped axisymmetric electric fields [9, 10]. Passive DEP levitation has the advantages of simplicity; the nonuniform electrostatic field holds the particle in static equilibrium, and no particle position detection or feedback control is required [1]. In order to levitate a particle exhibiting positive DEP (feedback controlled levitation) ($\mathrm{Re}[\underline{K}] > 0$) [11, 12], the gradient of a field is required, where an intensity maximum exists at an isolated point in the space away from the electrode surfaces. Such maxima do not occur in divergence and curl-free electrostatic fields [1].

The conditions required for the stable levitation of a particle experiencing a positive DEP force have been discussed at length by Kraybill [13]. The particle, assumed to be in equilibrium on the axis in an axisymmetric electric field, must be stable with respect to the radial and axial motions. Radial stabilization is achieved using a focused electric field with the intensity decreasing as the particle moves away from the axis. In such an electrode structure, sensing the vertical particle displacements from an equilibrium point and adjusting the levitation voltage to maintain the particle at the preset equilibrium point achieves stabilization in the axial direction. Two theories, dipole and quadrupole, are used to analyze DEP levitation.

3.1.2.4 Dipole Levitation Theory

A common configuration used to implement a dipole levitation system is the cone-plate assembly of Figure 3.6 whereby a rounded cone electrode is brought into close proximity to a conducting ground plane. This structure is derived from the analytical solution for the electrostatic potential due to a semi-infinite line charge over a ground plane (Figure 3.7), which can be shown [1] to be:

$$V = \frac{\lambda}{4\pi\varepsilon}\ln\left[\frac{h+z+\sqrt{(h+z)^2+\rho^2}}{h-z+\sqrt{(h-z)^2+\rho^2}}\right] \tag{3.18}$$

where λ is the line charge density in units of coulomb/meter.

The equipotential surfaces of this scalar field take the form of rounded cones and define the curvature of the conducting tip used for experimentation.

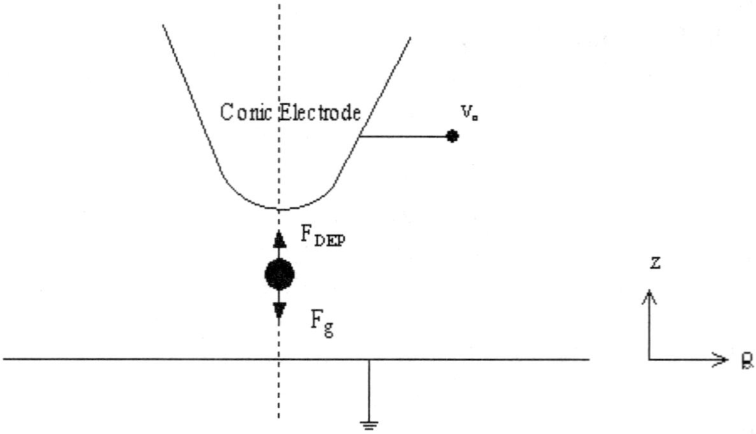

Figure 3.6 Cone-plate electrode structure.

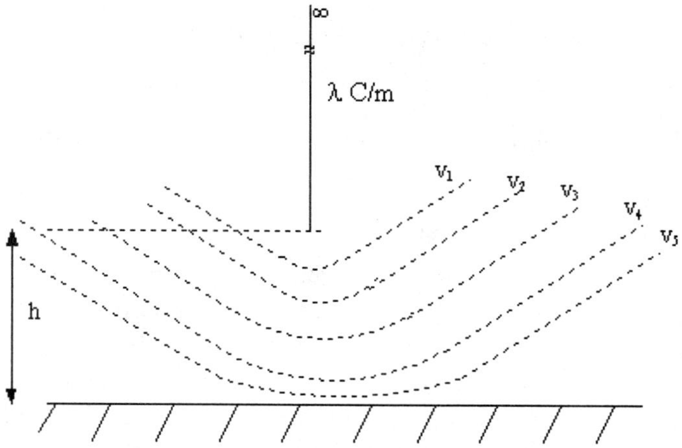

Figure 3.7 Semi-infinite line charge over ground plane.

Also, from the voltage solution of (3.18), the electric field along the z-axis is given by

$$E_z\Big|_{\rho=0} = -\frac{\partial V}{\partial z}\Big|_{\rho=0} \tag{3.19}$$

The gradient of its square can easily be shown to be

$$\begin{aligned}\nabla E_z^2 &= \frac{\partial E_z^2}{\partial z} = \frac{\lambda^2}{2\pi^2\varepsilon_m^2}\frac{2h^2 z}{\left(h^2 - z^2\right)^3} \\ &= \frac{\lambda^2}{2\pi^2\varepsilon_m^2}G_\infty(z)\end{aligned} \tag{3.20}$$

where the $G_\infty(z)$ collects the geometric dependencies, $\frac{2h^2 z}{(h^2 - z^2)^3}$. From (3.20), the ∇E^2 term in the dipolar dielectrophoretic force expression of (3.10) is now analytically defined:

$$\left\langle \vec{F}_{DEP} \right\rangle = 2\pi a^3 \varepsilon_m \operatorname{Re}[\underline{K}]\frac{\lambda}{2\pi^2\varepsilon_m^2}G_\infty(z) \tag{3.21}$$

$$\left\langle \vec{F}_{DEP} \right\rangle = \frac{\lambda^2}{\pi \varepsilon_m} a^3 \operatorname{Re}[K_1] G_\infty(z) \tag{3.22}$$

Therefore the DEP force $\left\langle \vec{F}_{DEP} \right\rangle$ is directly related to the (particle radius)^3 of the levitated particle.

3.1.2.5 Quadrupole Levitation Theory

The dipole approximation is not accurate when considering field-induced interactions in highly nonhomogenous fields because of induced higher-order field interactions [13]. Such cases can be handled readily by the multipolar theory [13–15]. In this theory, Legendre polynomial expansion in spherical coordinates of the potential due to an azimuthally symmetric arrangement of point charges is used. The model used for analytical representation of quadrupole behavior comprises four point charges (two $+Q$ and two $-Q$) arranged as shown in Figure 3.8. From the generalized expression in spherical coordinates for the potential (Φ_p) at some point (r, θ, ϕ) due to a point charge Q located at (r', θ', ϕ'), superposition is applied to determine the electrostatic potential in the vicinity of the quadrupole. A mathematical extension to the equivalent dipole method, the equivalent multipolar moment method [1] is used to determine the force on a vanishingly small sphere with surface charge σ_t on its surface with permittivity ε_1 and suspended in a medium of permittivity ε_o

$$F_z = \int_{surface} \frac{\varepsilon_1}{\varepsilon_o} \sigma_t E_z dS \tag{3.23}$$

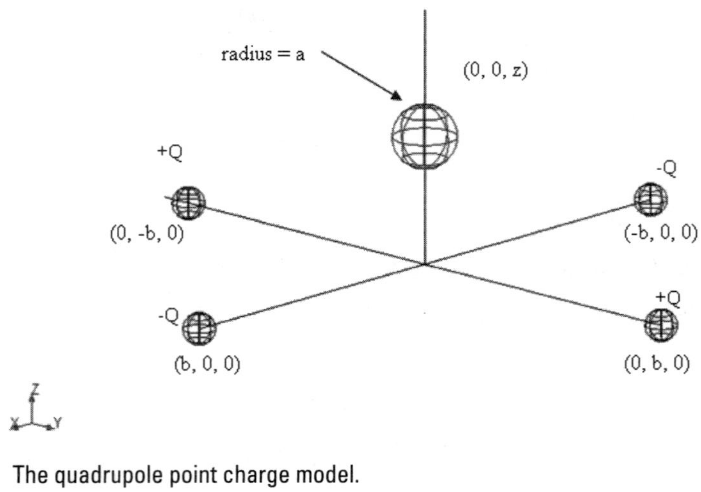

Figure 3.8 The quadrupole point charge model.

where the z-component of the electric field (E_z) is obtained from the axial derivative of the potential function. For particles of radius $a << \sqrt{b^2 + z^2}$, the series expression of the z-directed force component can be truncated and normalized by a^5 to become [2, 6]:

$$\frac{F_z}{a^5} \cong -\frac{3Q^2}{\pi\varepsilon_1}\mathrm{Re}[K_2]\frac{(z/b)}{b^7\left(1+(z/b)^2\right)^6} = -\frac{3Q^2}{\pi\varepsilon_1}\mathrm{Re}[K_2]G_{QUAD}(z) \tag{3.24}$$

where $G_{QUAD}(z)$ collects the geometric dependencies and

$$K_n = \frac{n(2n+1)\left(\varepsilon_p^* - \varepsilon_m^*\right)}{n\varepsilon_p^* + (n+1)\varepsilon_m^*} \tag{3.25}$$

Then K_2 in (3.25) will be:

$$K_2 = \frac{10\left(\varepsilon_p^* - \varepsilon_m^*\right)}{2\varepsilon_p^* + 3\varepsilon_m^*} \tag{3.26}$$

Contrasting the quadrupole levitation force on a sphere in (3.24) with the dipolar dielectrophoretic force on a sphere in (3.21), two fundamental differences are evident. First, the "radius cubed" dependence in the dipolar force expression is replaced by the "radius to the fifth" dependence in the quadrupole case. Second, the Claussius-Mosotti factor present in the dipole expression is replaced in the quadrupole expression by the higher order K_2 term. The quadrupolar force's dependency on (particle radius)^5 is the distinguishing feature that makes the quadrupole an attractive component for inclusion in particle processing systems.

3.2 Magnetic Field

Magnetic fields surround magnetic materials and electric currents and are detected by the force they exert on other magnetic materials and moving electric charges.

3.2.1 Nonuniform Magnetic Field (Magnetophoresis)

Magnetophoresis (MAP) [1] is defined as the motion of uncharged (neutral) particles caused by the polarization effect in a nonuniform magnetic field. Much interpretation and analysis of dielectrophoresis (DEP) carries over to magnetophoresis (MAP). However, most magnetic particles are ferromagnetic and exhibit strong nonlinearity, limiting the usefulness of the DEP-MAP analogy [1]. The features of the MAP phenomenon can be summarized as follows:

- Particles experience MAP force only when the magnetic field is nonuniform.
- The MAP force does not depend on the polarity of the applied magnetic field and is observed with AC as well as DC excitation.
- There are two kinds of MAP forces:
  - Positive MAP for $\mu_m < \mu_p$, where μ_m is the permeability of the suspended medium, and μ_p is the permeability of the particles. In this case, particles are attracted to regions of the stronger magnetic field.
  - Negative MAP for $\mu_m > \mu_p$. In this case, particles are repelled from regions of the stronger magnetic field.

3.2.2 Magnetophoresis Force (MAP Force)

Similarly to DEP force (Section 3.1.1), if we have a neutral particle suspended in some dielectric fluid and polarized by a nonuniform magnetic field $\vec{H}$, and due to magnetic polarization, $\vec{H}$ will induce a moment in the particle and a net magnetic force will be given as:

$$\overrightarrow{F_{MAP}} = \left(\vec{m}_{eff}.\nabla\right)\vec{H} \tag{3.27}$$

where $\vec{m}_{eff}$ is the effective magnetic dipole moment, and is defined as the moment of an equivalent, free charge, point dipole that when immersed in the same dielectric fluid and positioned at the same location as the center of the original particle, produces the same dipolar magnetic effect.

In the case of a sphere in a medium, both of which are lossless dielectrics, that is, both materials have finite dielectric constant and zero conductivity (ideal case), the induced effective dipole moment $\vec{m}_{eff}$ is given by [1].

$$\vec{m}_{eff} = 4\pi a^3 \left(\frac{\mu_p - \mu_m}{\mu_p + 2\mu_m}\right)\vec{H} \tag{3.28}$$

where a is the radius of the sphere, μ_m is the permeability of the medium, μ_p is the permeability of the sphere, and $\vec{H}$ is the strength of the applied magnetic field. Substituting (3.8) into (3.7), the net force on the sphere is:

$$\overrightarrow{F_{MAP}} = 2\pi a^3 \mu_m \left(\frac{\mu_p - \mu_m}{\mu_p + 2\mu_m} \right) \nabla \left| \vec{H} \right|^2 \tag{3.29}$$

The term $\left(\frac{\mu_p - \mu_m}{\mu_p + 2\mu_m} \right)$ is known as the Clausius-Mossotti factor (K) [1].

In the nonideal case, where ohmic losses are taken into consideration, a homogenous spherical particle with radius a and complex permittivity ω_{cp}, when immersed in a medium of complex permittivity ε_m^* and exposed to a spatially nonuniform AC electric field, the time average MAP force $\langle \vec{F}_{MAP} \rangle$ is defined as follows:

$$\langle \vec{F}_{MAP} \rangle = 2\pi a^3 \mu_m \operatorname{Re}[\underline{K}] \nabla H_{rms}^2 \tag{3.30}$$

where

$$\underline{K} = \left(\frac{\mu_p^* - \mu_m^*}{\mu_p^* + 2\mu_m^*} \right) \tag{3.31}$$

is the complex Clausius-Mossotti factor, and μ_p^* and μ_m^* are defined by (3.32) and (3.33), respectively:

$$\mu_p^* = \mu_p - j\frac{\sigma_p}{\omega} \tag{3.32}$$

$$\mu_m^* = \mu_m - j\frac{\sigma_m}{\omega} \tag{3.33}$$

where σ_p and σ_m are the conductivity of the particle and the medium, respectively, and ω is the angular frequency of the applied electric field. From (3.30), we can observe the following:

- The MAP force is linearly related to the volume of the immersed particle (i.e., $F_{MAP} \alpha a^3$).
- F_{MAP} is related directly to the permeability of the suspended medium (i.e., $F_{MAP} \alpha \mu_m$).
- The MAP force depends upon the magnitude and sign of the Clausius-Mossotti factor.
- The MAP force vector is directed along the gradient of the magnetic field intensity ∇H_{rms}^2, and the time-averaged MAP force is independent of the field polarity, as we are taking the rms value of the intensity of the magnetic field.

3.2.2.1 Positive and Negative MAP Forces

In accordance with (3.30), the direction of the MAP force can be determined by the real part of the Clausius-Mossotti factor ($\mathrm{Re}[\underline{K}]$). When ($\mathrm{Re}[\underline{K}]$) is positive (i.e., $\mu_m < \mu_p$), particles are attracted to magnetic field intensity maxima and repelled from the magnetic field intensity minima. This force is known as *positive magnetophoresis. Negative magnetophoresis* occurs when $\mathrm{Re}[\underline{K}] < 0$ (i.e., $\mu_m > \mu_p$), where particles are attracted to the magnetic field intensity minima and repelled from the magnetic field intensity maxima.

References

[1] Jones, T. B., *Electromechanics of Particles*, Cambridge, U.K.: Cambridge University Press, 1995.

[2] Dukhin, S. S., and B. V. Derjaguin, *Electrokinetic Phenomena*, New York: John Wiley & Sons, 1974.

[3] Pohl, H. A., *Dielectrophoresis*, Cambridge, U.K.: Cambridge University Press, 1978.

[4] Voldman, J., "A Microfabricated Dielectrophoretic Trapping Array for Cell-Based Biological Assays," Ph.D. thesis, Massachusetts Institute of Technology, June 2001.

[5] Suehiro, J., and R. Pethig, "The Dielectrophoretic Movement and Positioning of a Biological Cell Using a Three-Dimensional Grid Electrode System," *Journal of Physics D (Applied Physics)*, Vol. 27, No. 22, 1994, pp. 1571–1574.

[6] Kaler, K. V. I. S., and H. A. Pohl, "Dynamic Dielectrophoretic Levitation of Living Individual Cells," *IEEE Trans. on Industry Applications*, Vol. 6, No.19, 1983, pp. 1089–1093.

[7] Washizu, M., and T. B. Jones, "Equilibria and Dynamics of DEP-Levitated Particles: Multipolar Theory," *Journal of Electrostatics*, Vol. 33, 1994, pp. 199–212.

[8] Wang, X., et al., "Dielectrophortic Manipulation of Particles," *IEEE Trans. on Industry Applications*, Vol. 33, No. 3, 1997, pp. 660–669.

[9] Jones, T. B., and G. W. Bliss, "Bubble Dielectrophoresis," *Journal of Physics D (Applied Physics)*, Vol. 48, 1977, pp. 1414–1417.

[10] Jones, T. B., and J. P. Kraybill, "Active Feedback-Controlled Dielectrophoretic Levitation," *Journal of Physics D (Applied Physics)*, Vol. 60, 1986, pp. 1247–1252.

[11] Kraybill, J., "Digital Feedback Systems in Dielectrophoresis," Ph.D. thesis, University of Rochester, Rochester, NY, 1986.

[12] Washizu, M., and T. B. Jones, "Multipolar Dielectrophoretic Force Calculation," *Journal of Electrostatics*, Vol. 33, 1994, pp. 187–198.

[13] Washizu, M., T. B. Jones, and K. V. I. S. Kaler, "Higher-Order Dielectrophoretic Field Effects: Levitation at a Field Null," *Biochimica et Biophysica Acta*, Vol. 1158, 1993, pp. 40–46.

[14] Washizu, M., and T. B. Jones, "Equilibria and Dynamics of DEP-Levitated Particles: Multipolar Theory," *Journal of Electrostatics*, Vol. 33, 1994, pp. 199–212.

[15] Hunter, R. J., *Foundations of Colloid Science*, Oxford, U.K.: Oxford University Press, 1989.

4

Metal-Oxide Semiconductor (MOS) Technology Fundamentals

Semiconductors have been used to denote materials with a much higher conductivity than insulators; however, semiconductors have a much lower conductivity than metals measured at room temperature. Another factor that distinguishes between metals and semiconductors is the temperature dependence of the conductivity. At high temperatures, both metals and semiconductors retain their conductivity, while at low temperatures, semiconductors are transformed into insulators. Metals have a large carrier density; semiconductors exhibit a moderate carrier density at room temperature, while insulators have a negligible carrier density.

4.1 Semiconductor Properties

Semiconductors are the most important materials for the fabrication of electronic and optoelectronic devices. Silicon (Si) is the most common semiconductor material used for semiconductor devices and integrated circuits [1]. Other semiconductor materials are used for specialized applications, such as gallium arsenide (GaAs) and related compounds are used for high-speed devices and optical devices.

4.2 Intrinsic Semiconductors

The atom is the basic unit of matter consisting of a dense, central nucleus surrounded by a cloud of negatively charged electrons. The atomic nucleus contains a mix of positively charged protons and electrically neutral neutrons. The electrons are distributed in various shells at different distances from the nucleus. Electron energy increases as the shell radius increases. Electrons in the outermost shell are called valence electrons. The number of valence electrons identifies the chemical activity of a material [1].

Table 4.1 shows a portion of a periodic table. The common semiconductors such as silicon (Si) and germanium (Ge) are in group IV (elemental semiconductors), while gallium arsenide (compound semiconductor) is a group III-V [2].

The silicon crystal structure is a tetrahedral configuration in which each silicon atom has four nearest neighbors, as shown in Figure 4.1. The five noninteracting silicon atoms, with the four valence electrons of each atom, are emanating

Table 4.1
Periodic Table

III	IV	V
B	C	—
Al	Si	P
Ga	Ge	As

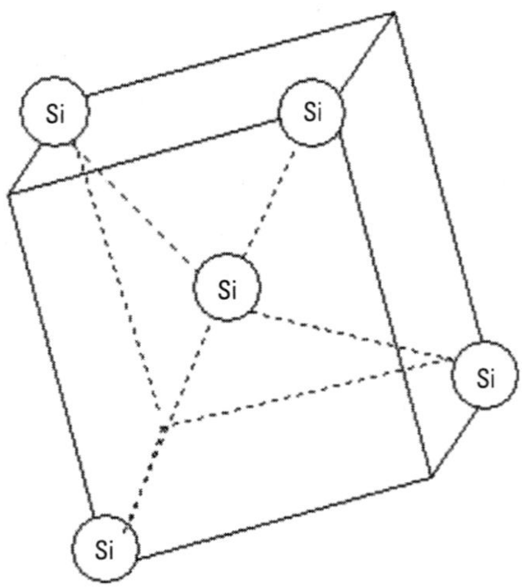

Figure 4.1 Five silicon atoms, each with four valance electrons.

from the atom. When silicon atoms come closer to each other, the valance electrons interact to form a crystal. The valance electrons are shared between atoms and form covalent bonds. Other semiconductor materials, such as germanium and gallium arsenide, have the same tetrahedral structure. In its lattice form, the valance electrons are always available on the outer edge of the silicon crystal; thus, additional atoms can be added to form large single crystal structures [1].

At zero Kelvin temperature ($T = 0$K), each electron is in its lowest energy state and the covalent bond positions are filled by electrons. The electrons will not move if an electric field is applied to this material, because they are still bound to their own atoms. Thus, at $T = 0$K, the silicon is an insulator (i.e., no charge flows through the silicon material). If the temperature increases, the valance electrons will gain thermal energy to break the covalent bonds and move away from their original positions; these electrons are free to move within the crystal. The net charge in the crystal is neutral (i.e., the number of positive charges is equal the negative charges). If a negatively charged electron breaks its covalent bond and moves away from its position, a positively charged "empty state" is created at the position. As the temperature increases, more covalent bonds are broken and more free electrons and empty states are created. The valance electron requires a minimum energy to break the covalent bond; this minimum energy is called E_g (bandgap energy). The energy gap E_g is higher in insulator compared to conductors, which contain very large numbers of free electrons at room temperature. In a semiconductor, E_g is in the order of 1 electron volt (eV). A current flow in a semiconductor occurs as a result of a net flow of free electrons [2]. In addition, a valance electron, which has a certain thermal energy and is adjacent to an empty state, may move into the position as shown in Figure 4.2. This movement can appear as a positive charge moves through the semiconductor. This positive charge is called a hole. In semiconductors, there are two types of charged particles that

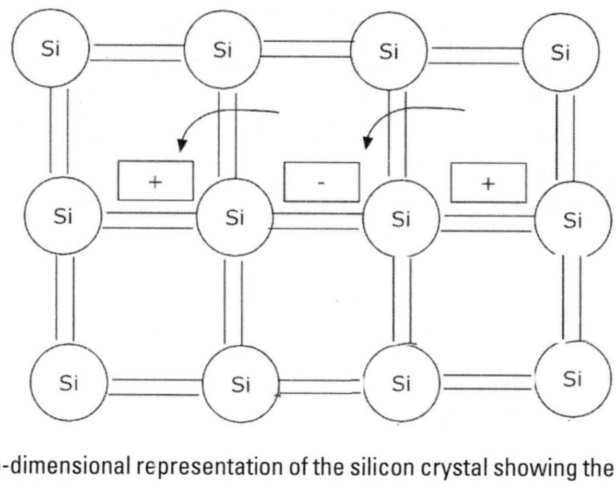

Figure 4.2 Two-dimensional representation of the silicon crystal showing the movement of the positively charged hole.

contribute to the current: the negatively charged free electrons and the positively charged holes. The magnitude of the current in the semiconductor material is directly influenced by the concentrations of the electrons and holes [2].

An intrinsic semiconductor is a single crystal semiconductor material with no other types of atoms within the crystal. Because the thermal energy is the only source for generating free electrons and holes in the intrinsic semiconductor, the densities of electrons and holes are equal [3]. The intrinsic carrier concentration (n_i) represents the concentration of free electrons as well as the holes. The equation for n_i is

$$n_i = BT^{\frac{3}{2}} e^{\left(\frac{-E_g}{2KT}\right)} \tag{4.1}$$

where B is a constant related to a specific semiconductor material, E_g is the bandgap energy (eV), T is the temperature (K), and k is the Boltzmann's constant (86 × 10^{-6} eV/ K). The values of E_g and B for different semiconductors are given in Table 4.2 [2].

Also, a pure semiconductor is often called an "intrinsic" semiconductor. The conductivity, or ability to conduct, of common semiconductor materials can be drastically changed by adding other elements, called "impurities" to the melted intrinsic material and then allowing the melt to solidify into a new and different crystal. This process is called "doping" [2].

4.3 Extrinsic Semiconductor

The electron and holes concentrations in an intrinsic semiconductor are small. Thus, very small hole or electron currents are possible. These currents can be increased by adding a controlled amount of certain impurities. Impurity enters the crystal lattice and replaces one of the semiconductor atoms, even though the impurity atom does not have the same valance electron structure. For silicon, the desirable substitutional impurities are from the group III and V elements (refer to Table 4.1). Phosphor and arsenic are the most common elements in group V. Figure 4.3 shows a phosphorous atom substituting a silicon atom, and four

Table 4.2
Semiconductor Constants

Material	B (cm–3 K–3/2)	Eg (eV)
Si	5.23×10^{15}	1.1
Ge	1.66×10^{15}	0.66
GaAs	2.1×10^{14}	1.4

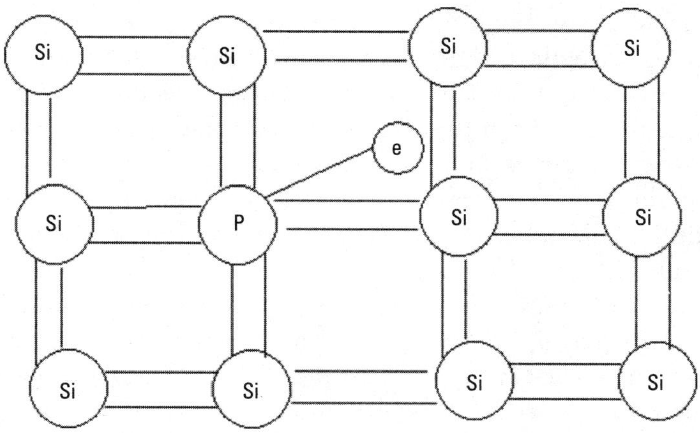

Figure 4.3 Two-dimensional representation of the silicon lattice doped with a phosphorous atom.

valance electrons of the phosphorous atom are used to satisfy the covalent bond requirements [4]. The fifth valance electron is loosely bound to the phosphorous atom. At room temperature, this electron has enough thermal energy to break the bond, thus being free to move through the crystal and contribute to the electron current in the semiconductor. Because the phosphorous atom donates an electron that is free to move, it is called a donor impurity [2]. The remaining phosphorous atom is immobile and cannot contribute to the current, although, the remaining phosphorous atom has a net positive charge. Therefore, when a donor impurity is added to a semiconductor, free electrons are created without generating holes. This process (known as the doping process) allows the concentration of free electrons in a semiconductor to be controlled as required.

In other words, an extrinsic semiconductor is a semiconductor that has been doped, that is, into which a doping agent has been introduced, giving it different electrical properties than the intrinsic (pure) semiconductor. Doping involves adding dopant atoms to an intrinsic semiconductor, which changes the electron and hole carrier concentrations of the semiconductor at the thermal equilibrium. Dominant carrier concentrations in an extrinsic semiconductor classify it as either an n-type or p-type semiconductor. The electrical properties of extrinsic semiconductors make them essential components of many electronic devices [2]. Semiconductor doping is the process that changes an intrinsic semiconductor to an extrinsic semiconductor. During doping, impurity atoms are introduced to an intrinsic semiconductor. Impurity atoms are atoms of a different element than the atoms of the intrinsic semiconductor. Impurity atoms act as either donors or acceptors to the intrinsic semiconductor, changing the electron and hole concentrations of the semiconductor. Impurity atoms are classified as donor or acceptor atoms based on the effect they have on the intrinsic semiconductor. Donor impurity atoms have more valence electrons than the atoms they replace in the intrinsic semiconductor lattice. Donor impurities "donate" their extra valence electrons to a semiconductor's conduction band, providing excess electrons to the intrinsic semiconductor. Excess electrons increase the electron carrier concentration (n_0) of the semiconductor, making it n-type. Acceptor impurity atoms have fewer valence electrons than the atoms they replace in the intrinsic semiconductor. They "accept" electrons from the semiconductor's valence band. This provides excess holes to the intrinsic semiconductor. Excess holes increase the hole carrier concentration (p_0) of the semiconductor, creating a p-type semiconductor. Semiconductors and dopant atoms are defined by the column of the periodic table of elements they fall in. The column definition of the semiconductor determines how many valence electrons its atoms have and whether dopant atoms act as the semiconductor's donors or acceptors. Group IV semiconductors use group V atoms as donors and group III atoms as acceptors. Group III-V semiconductors use group VI atoms as donors and group II atoms as acceptors. Group III-V semiconductors can also use group IV atoms as either donors or acceptors. When a group IV atom replaces the group III element in the semiconductor lattice the group IV atom acts as a donor. Conversely, when a group IV atom replaces the group V element the group IV atom acts as an acceptor. Group IV atoms can act as both donors and acceptors; therefore, they are known as amphoteric impurities [5].

4.3.1 N-Type Doping

Extrinsic semiconductors with a larger electron concentration than hole concentration are known as n-type semiconductors. The phrase "n-type" comes from the negative charge of the electron [2]. In n-type semiconductors, electrons are the majority carriers and holes are the minority carriers. N-type semiconductors are created by doping an intrinsic semiconductor with donor impurities. In an n-type semiconductor, the Fermi energy level is greater than that of the intrinsic semiconductor and lies closer to the conduction band than the valence band.

4.3.2 P-Type Doping

P-type semiconductors, in comparison to n-type, have a larger hole concentration than electron concentration. The phrase "p-type" refers to the positive charge of the hole. In p-type semiconductors, holes are the majority carriers and electrons are the minority carriers. P-type semiconductors are created by doping an intrinsic semiconductor with acceptor impurities. P-type semiconductors have Fermi energy levels below the intrinsic Fermi energy level. The Fermi energy level lies closer to the valence band than the conduction band in a p-type semiconductor [1].

4.4 MOS Device Physics

A metal-oxide semiconductor (MOS) structure is obtained by growing a layer of silicon dioxide (SiO_2) on top of a silicon substrate and depositing a layer of metal or polycrystalline silicon (the latter is commonly used). As the silicon dioxide is a dielectric material, its structure is equivalent to a planar capacitor, with one of the electrodes replaced by a semiconductor [1]. Figure 4.4 shows the MOS capacitor. The physics of the MOS structure can be explained using a simple parallel plate capacitor, as shown in Figure 4.5. The top plate has a negative voltage with respect to the bottom plate, and an insulator material separates the two plates. According to the bias voltage, negative and positive charges exist in the top and bottom plates, respectively, as shown in Figure 4.5.

A MOS capacitor with a p-type semiconductor substrate is shown in Figure 4.6. The top metal terminal is called the gate, which is at a negative voltage with respect to the semiconductor substrate [2]. The direction of the electric field is as shown in Figure 4.6.

If the electric field penetrates the substrate, the holes in the p-type substrate will experience a force toward the oxide-semiconductor interface. Accordingly, the holes will accumulate at the bottom plate of the MOS capacitor. If

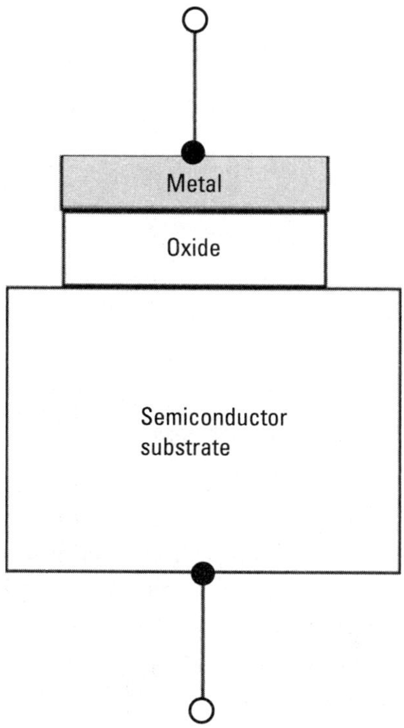

Figure 4.4 The MOS capacitor structure.

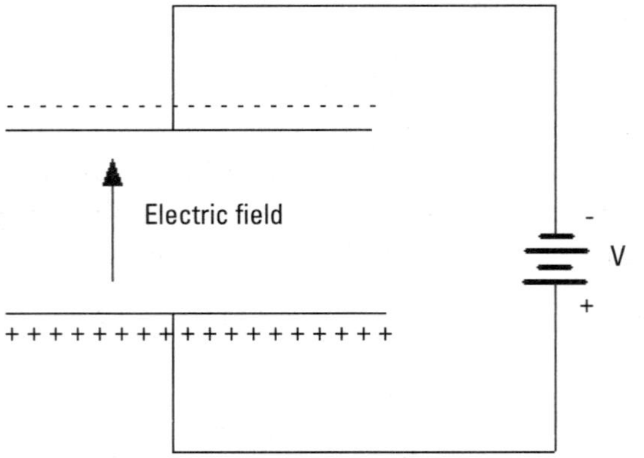

Figure 4.5 A parallel-plate capacitor shows the charges and the electric field.

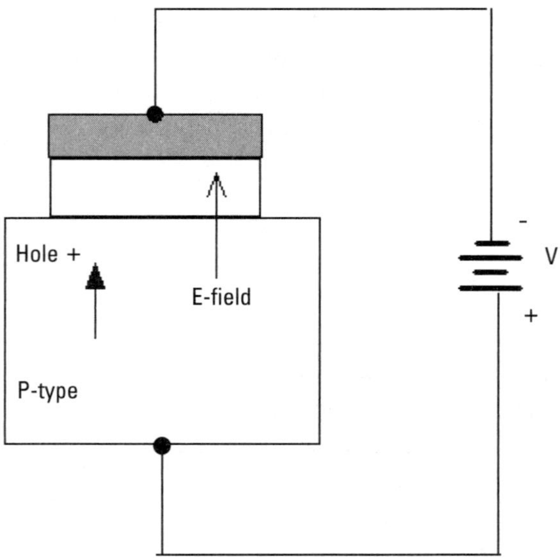

Figure 4.6 MOS capacitor with a negative gate bias.

the polarity of the applied voltage is reversed, a positive charge will exist on the top of the metal plate and the induced electric field is in the opposite direction, as shown in Figure 4.7. The holes will be pushed away from the interface and a negative space charge region is created. The negative charge in the induced depletion region corresponds to the negative charge on the bottom plate of the MOS capacitor. The induced space charge region is shown in Figure 4.8. If the positive voltage applied to the gate is increased, the magnitude of the induced electric field will be increased. Minority carrier electrons are attracted to the

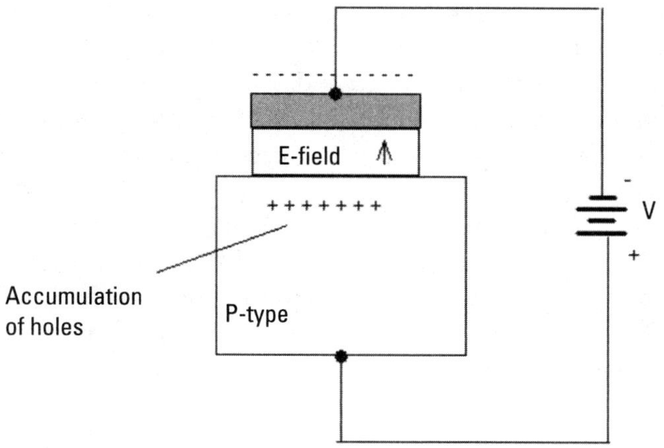

Figure 4.7 MOS capacitor with an accumulation layer of holes.

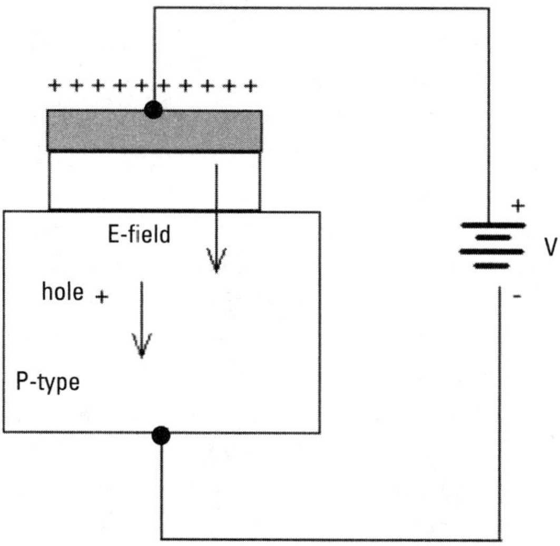

Figure 4.8 MOS capacitor with p-type substrate.

oxide-semiconductor interface, as shown in Figure 4.8. This region of the minority carrier electrons is called an electron inversion layer. The gate voltage controls the number of charges in the inversion layer [2]. By similarity, an n-type semiconductor substrate can provide the same basic charge distribution in a MOS capacitor. If the polarity of the applied voltage is reversed, a negative charge will exist on the top of the metal plate and the induced electric field is in the opposite direction. The electrons will be pushed away from the interface and a negative space charge region is created. The negative charge in the induced depletion region corresponds to the positive charge on the bottom plate of the MOS capacitor. If the positive voltage applied to the gate is increased, the magnitude of the induced electric field will be increased. Minority carrier holes are attracted to the oxide-semiconductor interface. This region of the minority carrier holes is called a holes inversion layer. The gate voltage controls the number of charges in the inversion layer.

The term *enhanced mode* means that a voltage must be applied to the gate to create an inversion layer [6]. For an n-type MOS capacitor, a negative voltage is required to be applied to create the electron inversion layer. While, for the p-type MOS capacitor, a positive voltage is required to be applied to create the electron inversion layer.

4.5 MOS Characteristics

A metal-oxide semiconductor field-effect transistor (MOSFET) is based on the modulation of a charge concentration by a MOS capacitance. The charges accumulate between a body electrode and a gate electrode located above the body and insulated from all other device regions by a gate dielectric layer, which in the case of a MOSFET is an oxide [6]. A metal-insulator semiconductor FET (MISFET) is a device that has a dielectric other than an oxide such as silicon dioxide [7]. Figure 4.9 shows the MOSFET structure. Compared to the MOS capacitor, the MOSFET includes two additional terminals (source and drain), each connected to individual highly doped regions that are separated by the body region [8]. These regions can be either p- or n-type, but they must both be of the same type, and of an opposite type to the body region. The source and drain (unlike the body) are highly doped as signified by a "+" sign after the type of doping.

If the MOSFET is a p-channel or p MOSFET, then the source and drain are p+ regions and the body is an n region. With sufficient gate voltage (refer to Section 4.4) above a threshold voltage value, electrons from the source enter the inversion layer or n-channel at the interface between the n region and the oxide [9]. This conducting channel extends between the source and the drain, and current is conducted through it when a voltage is applied between source and drain. Only a very small subthreshold leakage current can flow between the source and the drain. When the gate voltage is below the threshold value, the channel is lightly populated.

If the MOSFET is an n-channel or n MOSFET, then the source and drain are n+ regions and the body is a p region. When a positive gate-source voltage (negative source-gate) is applied, it creates an n-channel at the surface of the p region, analogous to the p-channel case, but with opposite polarities of charges and voltages [6, 10]. Only a very small subthreshold leakage current can flow between the source and the drain. When the gate voltage is below the threshold

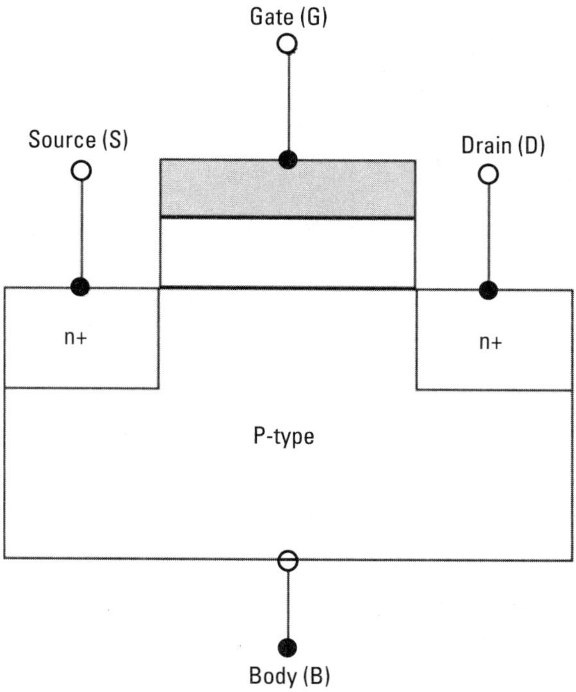

Figure 4.9 Cross-section of the n-channel MOSFET.

value, the channel is lightly populated. The source is so named because it is the source of the charge carriers (electrons for the n-channel, holes for the p-channel) that flow through the channel; similarly, the drain is where the charge carriers leave the channel.

4.5.1 Modes of Operation

The operation of a MOSFET can be separated into three different modes, depending on the voltages at the terminals. The threshold voltage (V_{th}) is defined as the applied gate voltage needed to create an inversion charge in the MOSFET channel. In other words, the minimum gate voltage required to turn on the MOSFET. For n-channel enhancement mode MOSFET, the threshold voltage is positive because a positive gate voltage is required to create the inversion charge [6, 11]. The three operational modes for the n-channel MOSFET are: cutoff (weak-inversion), linear, and saturation, as shown in Figure 4.10.

4.5.1.1 Cutoff (Weak-Inversion) Mode

When the voltage between the gate and source (V_{GS}) is less than threshold voltage (V_{th}) (i.e., $V_{GS} < V_{th}$), the current is considered to be zero, the transistor is

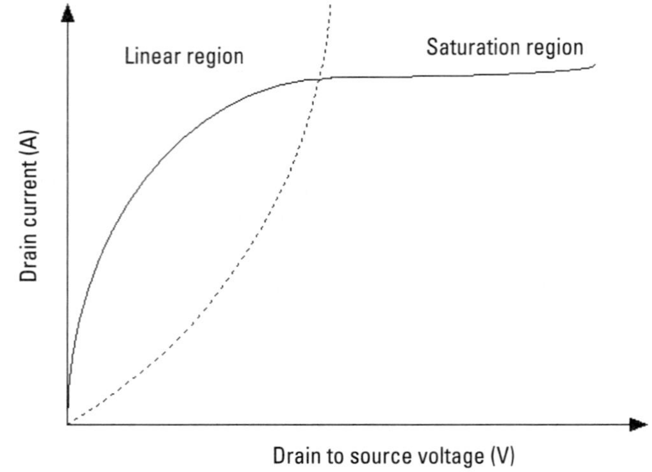

Figure 4.10 Current voltage characteristics of the MOSFET.

turned off, and there is no conduction between the drain and source [6]. In reality, some energetic electrons at the source enter the channel and flow to the drain, resulting in a subthreshold current that is an exponential function of gate-source voltage.

While the current between drain and source should ideally be zero when the transistor is being used as a turned-off switch, there is a weak-inversion current, sometimes called the subthreshold leakage current [12].

In weak inversion the current varies exponentially with gate-to-source bias V_{GS} as given approximately by [6]:

$$I_D \approx I_{D0} e^{\frac{V_{GS}-V_{th}}{nV_T}} \tag{4.2}$$

where I_{D0} = current at $V_{GS} = V_{th}$ and the slope factor n is given by

$$n+1+C_D/C_{OX} \tag{4.3}$$

where C_D is capacitance of the depletion layer and C_{OX} is capacitance of the oxide layer. In a long-channel device, there is no drain voltage dependence of the current once $V_{DS} >> V_{th}$, but as the channel length is reduced, drain-induced barrier lowering introduces drain voltage dependence that depends in a complex way upon the device geometry (for example, the channel doping, the junction doping, and so on). Frequently, threshold voltage V_{th} for this mode is defined as the gate voltage at which a selected value of current I_{D0} occurs, for example, I_{D0}

= 0.1 μA, which may not be the same V_{th} value used in the equations for the following modes.

By working in the weak-inversion region, some micropower MOSFET analog circuits are designed to take advantage of subthreshold conduction [6, 12]. These circuits deliver the highest possible transconductance-to-current ratio, namely: $g_m/I_D = 1/(nV_T)$. The subthreshold I-V curve depends exponentially upon threshold voltage, introducing a strong dependence on any manufacturing variation that affects threshold voltage, for example: variations in oxide thickness, junction depth, or body doping that change the degree of drain-induced barrier lowering. The resulting sensitivity to fabrication variations complicates optimization for leakage and performance [13, 14].

4.5.1.2 Triode (Linear) Mode

When the voltage between the gate and source (V_{GS}) is greater than threshold voltage (V_{th}) and the voltage between the drain and source V_{DS} is less than (V_{GS} − V_{th}) [i.e., $V_{GS} > V_{th}$ and $V_{DS} < (V_{GS} - V_{th})$] [6]. The transistor is turned on, and a channel has been created, which allows current to flow between the drain and the source. The MOSFET operates like a resistor, controlled by the gate voltage relative to both the source and drain voltages. The current from drain to source is modeled as:

$$I_D = \mu_n C_{ox} \frac{W}{L}\left((V_{GS} - V_{th})V_{DS} - \frac{V_{DS}^2}{2} \right) \tag{4.4}$$

where μ_n is the charge-carrier effective mobility, W is the gate width, L is the gate length, and C_{ox} is the gate oxide capacitance per unit area.

4.5.1.3 Saturation (Active) Mode

When the voltage between the gate and source (V_{GS}) is greater than the threshold voltage (V_{TH}), and the voltage between the drain and source V_{DS} is greater than (V_{GS} − V_{th}) (i.e., $V_{GS} > V_{th}$ and $V_{DS} > (V_{GS} - V_{th})$, the switch is turned on, and a channel has been created, which allows current to flow between the drain and source [2, 6, 15]. Since the drain voltage is higher than the gate voltage, the electrons spread out, and conduction does not occur through a narrow channel but through a broader, two- or three-dimensional current distribution extending away from the interface and deeper in the substrate. The onset of this region is also known as pinch-off to indicate the lack of channel region near the drain. The drain current is now weakly dependent upon drain voltage, controlled primarily by the gate-source voltage, and modeled very approximately as:

$$I_D = \mu_n C_{ox} \frac{W}{L}\left(\left(V_{GS} - V_{th}\right)2\left(1 + \lambda V_{DS}\right)\right) \tag{4.5}$$

The additional factor involving λ, the channel-length modulation parameter, models current dependence on drain voltage due to the Early effect, or channel length modulation [15]. According to this equation, a key design parameter, the MOSFET transconductance is:

$$g_m = \frac{2I_D}{V_{GS} - V_{th}} \tag{4.6}$$

where $V_{GS} - V_{th}$ is called the overdrive voltage [6, 15]. Another key design parameter is the MOSFET output resistance r_O given by:

$$r_o = \frac{1 + \lambda V_{DS}}{\lambda I_{DS}} \tag{4.7}$$

If λ is taken as zero, an infinite output resistance of the device results in unrealistic circuit predictions, particularly in analog circuits. As the channel length becomes very short, these equations become quite inaccurate. New physical effects arise. For example, carrier transport in the active mode may become limited by velocity saturation. When velocity saturation dominates, the saturation drain current is more linear than quadratic in V_{GS}. At even shorter lengths, carriers transport with near zero scattering, known as quasi-ballistic transport. In addition, the output current is affected by drain-induced barrier lowering of the threshold voltage.

4.6 Complementary Metal-Oxide Semiconductor (CMOS) Device

The principal reason for the success of the MOSFET was the development of digital complementary metal–oxide semiconductor (CMOS) logic, which uses p- and n-channel MOSFETs as building blocks [2, 6]. Overheating is a major concern in integrated circuits since increasingly more transistors are packed into ever smaller chips. CMOS logic reduces power consumption because no current flows (ideally), and thus no power is consumed, except when the inputs to logic gates are being switched. CMOS accomplishes this current reduction by complementing every n MOSFET with a p MOSFET and connecting both gates and both drains together. A high voltage on the gates will cause the n MOSFET to conduct and the p MOSFET not to conduct, and a low voltage on the gates causes the reverse. During the switching time, as the voltage goes from one state to another, both MOSFETs will conduct briefly. This arrangement greatly reduces power consumption and heat generation. Digital and analog CMOS applications are described in the following sections.

CMOS is a technology for constructing integrated circuits. CMOS technology is used in microprocessors, microcontrollers, static RAM, and other digital logic circuits. CMOS technology is also used for a wide variety of analog circuits such as image sensors, data converters, and highly integrated transceivers for many types of communication. Two important characteristics of CMOS devices are high noise immunity and low static power consumption. Significant power is only drawn while the transistors in the CMOS device are switching between on and off states. Consequently, CMOS devices do not produce as much waste heat as other forms of logic, for example transistor-transistor logic (TTL) or NMOS logic, which uses all n-channel devices without p-channel devices. CMOS also allows a high density of logic functions on a chip. It was primarily this reason why CMOS won the race and became the most used technology to be implemented in VLSI chips.

4.6.1 Advantages of CMOS Technology

There are two main advantages in using the CMOS technology. The first is signal quality. The signal is conditioned by means of dedicated circuitry units. This occurs perfectly at the electrode locations. The second advantage is integration. Large numbers of electrodes and transducers can be integrated on-chip and they can be addressed by on-chip multiplexing architectures. The use of CMOS offers large numbers of electrodes on a single chip, which provide good signal quality and improve the signal-to-noise ratio [16].

References

[1] Neman, D., *Electronic Circuit Analysis and Design*, Chicago, IL: Irwin, 1996.

[2] Neamen, D., *Semiconductor Physics and Devices: Basic Principles*, 3rd ed., New York: McGraw-Hill, 2003.

[3] Gray, P. R., et al., *Analysis and Design of Analog Integrated Circuits*, 4th ed., New York: John Wiley & Sons, 2001, pp. 66–67.

[4] van der Meer, P. R., A. van Staveren, and A. H. M. van Roermund, *Low-Power Deep Sub-Micron CMOS Logic: Subthreshold Current Reduction*, Dordrecht, the Netherlands: Springer, 2004.

[5] Smith, L. S., and A. Hamilton, *Neuromorphic Systems: Engineering Silicon from Neurobiology*, New York: World Scientific, 1998.

[6] Sedra, A. S., and K. C. Smith, *Microelectronic Circuits*, 5th ed., New York: Oxford, 2004.

[7] Vittoz, E. A., "The Fundamentals of Analog Micropower Design," in *Circuits and Systems Tutorials*, New York: John Wiley & Sons, 1996, pp. 365–372.

[8] Shukla, S. K., and R. Bahar, *Nano, Quantum, and Molecular Computing*, New York: Springer, 2004.

[9] Srivastava, A., D. Sylvester, and D. Blaauw, *Statistical Analysis and Optimization for VLSI: Timing and Power*, New York: Springer, 2005.

[10] Galup-Montoro, C., and M. C. Schneider, *MOSFET Modeling for Circuit Analysis and Design*, New York: World Scientific, 2007.

[11] Malik, N. R., *Electronic Circuits: Analysis, Simulation, and Design*, Englewood Cliffs, NJ: Prentice-Hall, 1995.

[12] Roy, K., and K. Seng Yeo, Low *Voltage, Low Power VLSI Subsystems*, New York: McGraw-Hill, 2004.

[13] Chen, W. K., *The VLSI Handbook*, Boca Raton, FL: CRC Press, 2006.

[14] Orshansky, M., S. Nassif, and D. Boning, *Design for Manufacturability and Statistical Design: A Constructive Approach*, New York: Springer, 2007.

[15] Colinge, J. P., and C. A. Colinge, *Physics of Semiconductor Devices*, Dordrecht, the Netherlands: Springer, 2002.

[16] Lee, H., D. Ham, and R. Westervelt, CMOS *Biotechnology*, New York: Springer, 2006.

5

Sensing Techniques for Lab-on-a-Chip

Many techniques have been used for sensing, analyzing, and monitoring the behavior of cells under the influence of electric or magnetic fields. They include optical techniques [1–3], fluorescent labeling [e.g., microfabricated fluorescence-activated cells sorter (μFACS)] technique [4–6], the impedance sensing technique [7–11], and the magnetic field sensing technique [12–14]. The sensing parts currently used in lab-on-a-chip are based on impedance sensing [15–17] and optical (photo detectors) techniques [18, 19].

5.1 Optical Technique

In this method, a dual beam optical spectrometer has been reported for rapid measurement of the dielectrophoretic behavior of cells and other particles subjected to a DEP force [1]. The optical system used is shown in Figure 5.1. It has an optical chamber, which incorporates two microelectrode arrays, between which cell suspensions are subjected to nonuniform electric fields and consequently DEP force.

Dielectrophoretically, the induced motion of the colloidal particles was monitored as a change in the optical density of the suspension by measuring the intensity of a light beam that passes through the gap between the two electrode arrays. Conditions of positive DEP, where the particles are removed from the bulk solution to collect at the edges of the electrodes, are thus monitored as a decrease in optical absorbance of the bulk solution. Negative DEP, where the particles are directed into the low-field regions in the bulk solutions away from the electrodes, is detected as an increase in the optical absorbance. The light from a laser diode is split into two parallel beams using a half and full-mirror as shown in Figure 5.1. The beam emerging from the half mirror is focused down by means of a standard microscope into the chamber containing the colloidal sample, while the other beam is similarly focused into a second (reference) chamber that contains fluid of the same composition as that used for the colloid suspending medium. This dual beam arrangement minimizes effects associated with fluctuations of light intensity and thus results in an improved signal-to-noise ratio compared with the other systems [2, 3].

The advantage of the optical technique is the ability to characterize the positive and negative DEP, which in turn provides a technique for deriving the practical conditions required for the selective manipulation and separation of cells using dielectrophoretic forces. The disadvantages of this technique from the lab-on-a-chip point of view can be summarized as follows: it requires bulky and expensive equipment, it needs complex sample preparation, and it is not suitable for miniaturization.

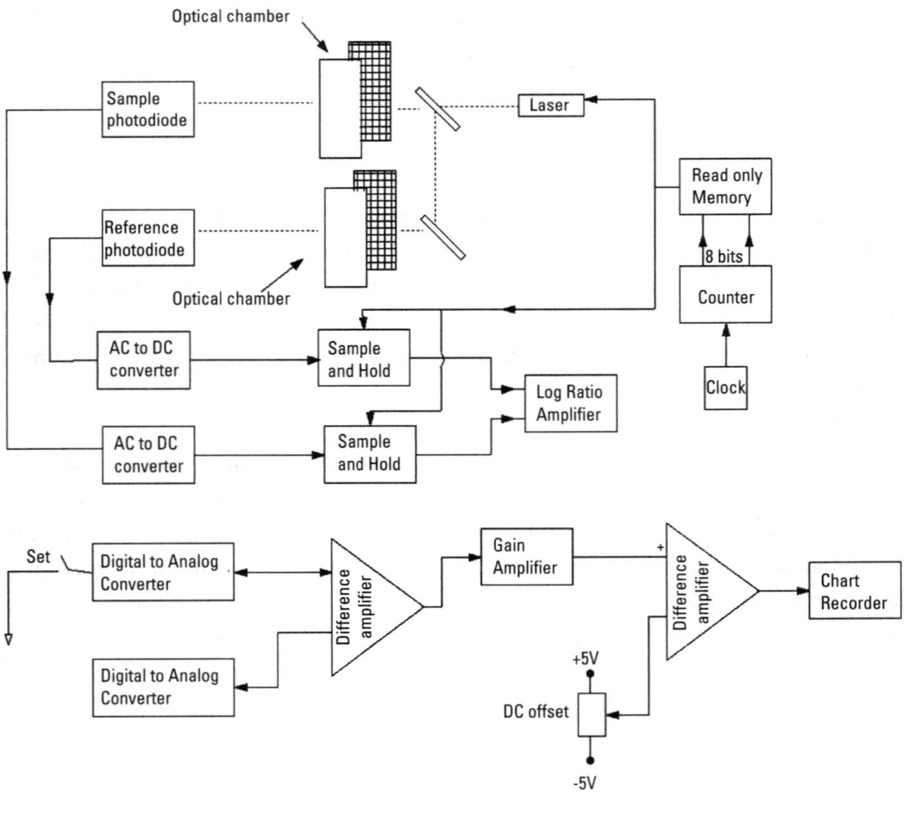

Figure 5.1 The overall blocks diagram of the dual DEP spectrometer [1].

5.2 Fluorescent Labeling Technique

Based on this technique, cell suspension containing cells labeled with a fluorescent dye have been used. The cells can be characterized and sorted by detecting the fluorescence of the cells [4]. In this technique, the fluorescence-activated cell sorter (FACS) machine is used [5]. This machine can rapidly separate the cells in a suspension on the basis of size and the color of their fluorescence; a schematic representation of the FACS is shown in Figure 5.2. The FACS machine works as follows: a cell suspension containing cells labeled with a fluorescent dye is directed into a thin stream so that all the cells pass in a single file. This stream emerges from a nozzle vibrating at some 40,000 cycles per second, which

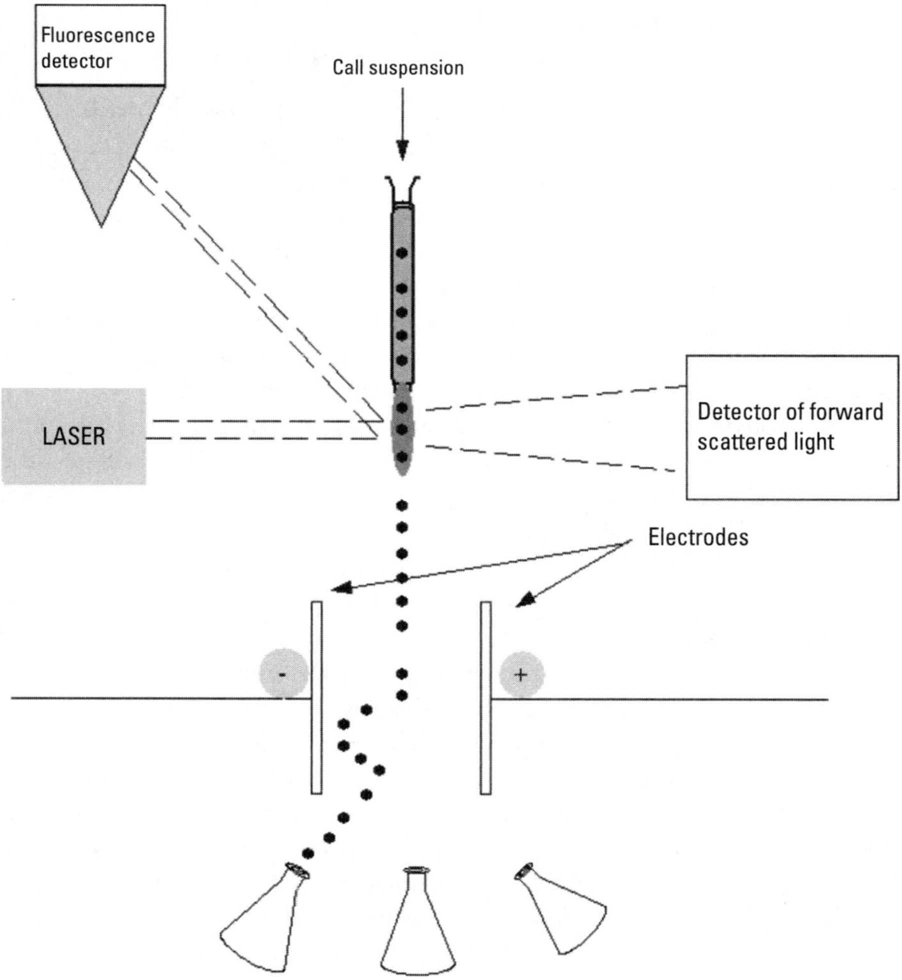

Figure 5.2 Schematic representation of the fluorescence-activated cell sorter (FACS) [6].

breaks the stream into 40,000 discrete droplets each second, some of which may contain a cell. A laser beam is directed at the stream just before it breaks up into droplets. As each labeled cell passes through the beam, its resulting fluorescence is detected by a photocell. If the signals from the two detectors (i.e., fluorescence and forward-scattered light detectors) meet either of the criteria set for fluorescence and size, an electrical charge (+ or −) is given to the stream. The droplets retain this charge as they pass between a pair of charged metal plates, such that:

- Positively charged drops are attracted to the negatively charged plate and vice versa.
- Uncharged droplets (those that contain no cell or a cell that fails to meet the desired criteria of fluorescence and size) pass straight into a third container and are later discarded.

The process does not damage the cells. In fact, because the machine can be set to ignore droplets containing dead cells, the percent viability of the sorted cells can be higher than that in the original suspension.

Although the FACS technique provides impressively efficient sorting, the devices are expensive and mechanically complex. Inexpensive devices that rapidly sort live cells, particles, and even single molecules would greatly facilitate screening of combinatorial chemistry libraries or cell populations during in vitro molecular evolution. Moreover, such devices would have wide applications in clinical medicine and basic biological and materials research.

A disposable microfabricated fluorescence-activated cell sorter (μFACS) was demonstrated in [6]. Compared with conventional FACS machines, the μFACS provides higher sensitivity, no cross-contamination, and a lower cost. The μFACS device is a silicone elastomer chip with three channels joined at a t-shaped junction (Figure 5.3). The channels are sealed with a glass cover slip. A buffer solution is introduced at the input channel and fills the device by capillary action. The pressure is equalized by adding a buffer to the two output ports and then adding a sample containing the cells to the input port. The cells are manipulated with electro-osmotic flow, which is controlled by three platinum electrodes at the input and output wells. The chip is mounted on an inverted optical microscope, and fluorescence is excited near the t-shaped junction with a focused laser beam. The fluorescent emission is collected by the microscope and measured with a photomultiplier tube (PMT). A computer digitizes the PMT signal and controls the flow by the electro-osmotic potentials (Figure 5.4).

The advantages of the fluorescent labeling technique are high sensitivity and efficient sorting. The disadvantages from the lab-on-a-chip point of view are that it requires cell modification by markers or antibody, and that the equip-

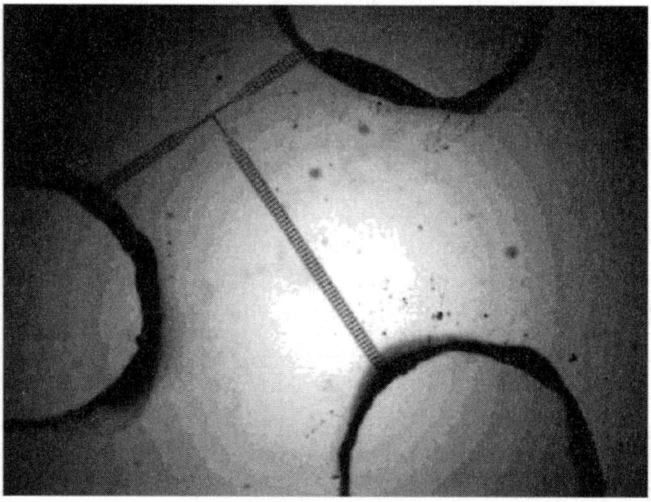

Figure 5.3 Optical micrograph of the μFACS device [6].

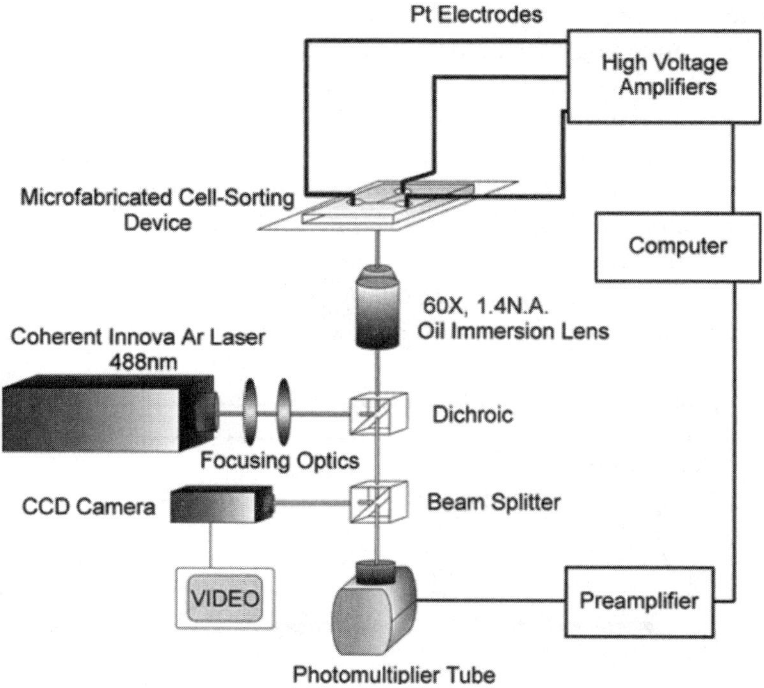

Figure 5.4 Schematic diagram of the cell sorting apparatus [6].

ment is rather expensive, bulky, and complex to operate. Thus it is not suitable for miniaturization.

5.3 Impedance Sensing Technique

It has been demonstrated that micromachining technologies may alleviate some of the problems encountered in macroscale flow cytometry [7]. On-chip integration of microfluidics and microelectrodes has already been reported for stationary cell measurements, trapping, and sorting [8]. The common attribute of impedance flow cytometry devices is that they are all able to measure the size of the cell under consideration. The measurement consists in determining the impedance change ΔR due to the particle passing through an aperture placed between two electrodes [9]. In a microchannel, the aperture is inbuilt and channel walls appear to be the best solution to achieve a high $\Delta R/R$ ratio by placing the electrodes as close as possible to each other and concentrating the current lines on the cell. The measurements are realized by sensing the differential variation of impedance $Z_{AC} - Z_{BC}$ into two successive channel segments as the cell passes consecutively into each one (Figure 5.5). This method is preferred over the use of a separate reference channel because the cell's properties can be measured directly against its surrounding media. Another advantage derives from the fact that the measurement and reference electrodes are inherently switched; any uneven drift of the electrode properties can be sensed and corrected, as both signal maxima should be of the same amplitude. Furthermore, the speed of the particle can be determined, as the distance between the two measurement areas and time t_{tr} separating both signal spikes are known, as shown in Figure 5.6.

An electric model is shown in Figure 5.7, where C_{dl} is the double-layer capacitance, R_c is the resistivity of the cell, C_m is the capacitance of the surrounding nonconductive membrane, and R_{sol1} is the resistance of the fluid surrounding

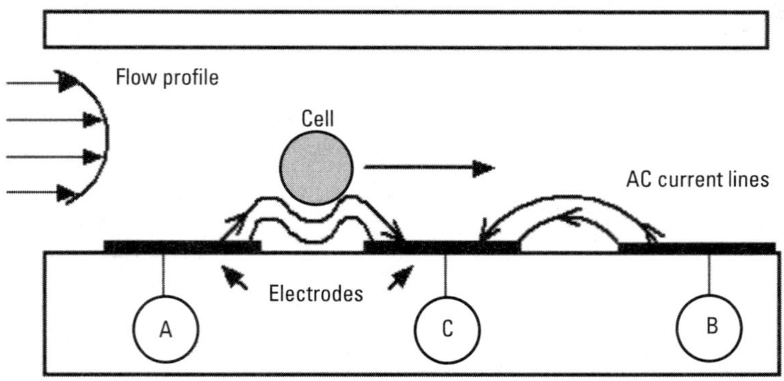

Figure 5.5 Side schematic view of the microchannel [10].

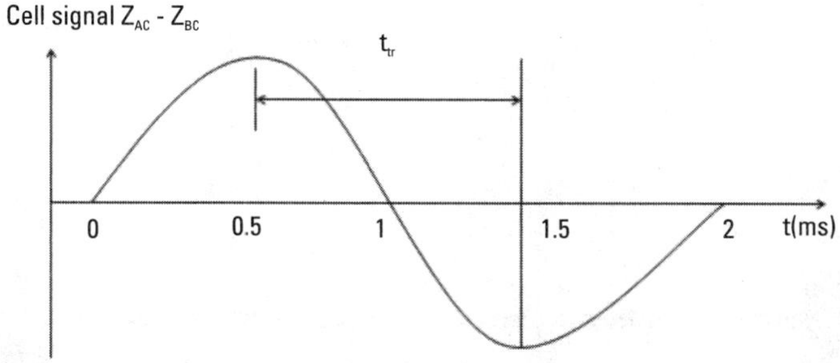

Figure 5.6 Impedance difference signal [10].

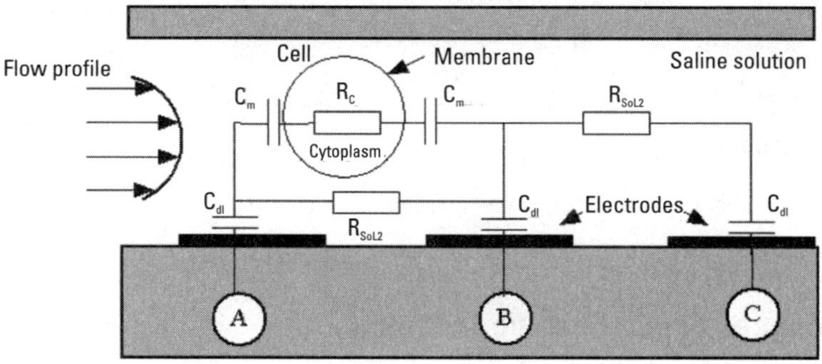

Figure 5.7 An electrical model of the impedance change [10].

the cell. Based on this model, the impedance change can be determined either analytically or through simulation using 3D finite elements [10].

Another method based on micromachining technology has been developed, the capacitance cytometry technique [11]. Using this technique, it is possible to quantify the DNA content of single eukaryotic cells from a diverse set of organisms, ranging from yeast to mammals. In addition, capacitance cytometry can be used as an assay for abnormal changes in DNA content, such as are encountered frequently in neoplastic cells. Because DNA is a highly charged molecule, in an applied low-frequency AC electric field its polarization response, in combination with the motion of the surrounding counter ions, can be substantial. The technique measures this response as a change in total capacitance, ΔC_T, across a pair of microelectrodes as individual eukaryotic cells suspended in a buffer solution flow one by one through a microfluidic channel, as shown in Figure 5.8.

The advantages of the impedance-sensing technique are that it can be used in many different tasks (e.g., counting, sizing, and population study), and that it is suitable for miniaturization. The disadvantage of this technique is that it does not provide integrated actuation capabilities and it requires microfluidic techniques to move cells in the device.

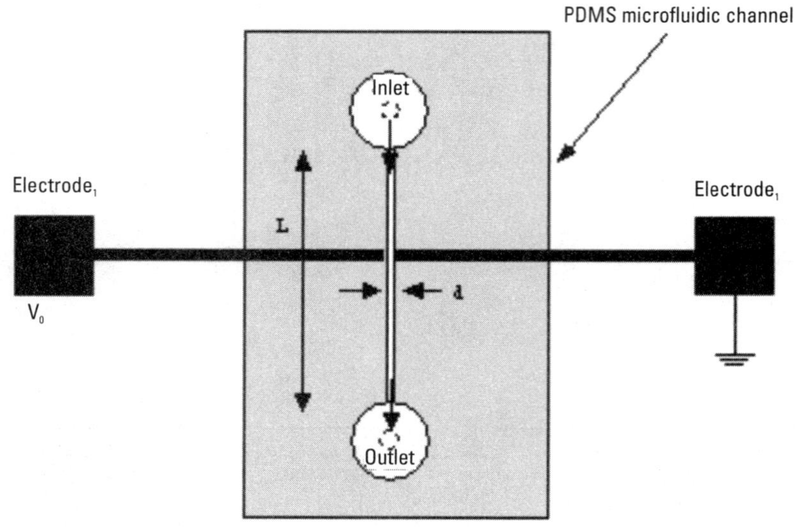

Figure 5.8 Schematic illustration of the microfluidic device [11].

5.4 Magnetic Field Sensing Technique

A magnetic field can control and manipulate biocells [20]. To sense the magnetic field, a magnetic field MOSFET (MAGFET) sensor is used [12–14]. MAGFET is essentially a MOSFET having one source and two drains as shown in Figure 5.9(a). The operation of the MAGFET structure is illustrated in Figure 5.9(b). In the absence of a magnetic field, the current flowing through the two drains of the MOSFET, I_{d1} and I_{d2}, are equal. When a magnetic field is applied, an unbalance in the current flowing through the two drains is produced by the Lorentz force [13]. This unbalance is a function of the intensity of the magnetic field applied. The disadvantages of the MAGFET are: it is a temperature-dependent device, and it has a large offset [14].

5.5 CMOS AC Electrokinetic Microparticle Analysis System

Keliman et al. presented the lexel array drive system [21]. This system consists of an array of lexel units, which is an electric field element. A lexel array is a programmable two-dimensional array of discrete, independent microelectrodes that constructs arbitrary electric field shapes to manipulate microscopic particles and cells. The lexel array forms the base structure of a future low-power bioanalysis platform. The analysis techniques employed by the lexel array exploit AC elec-

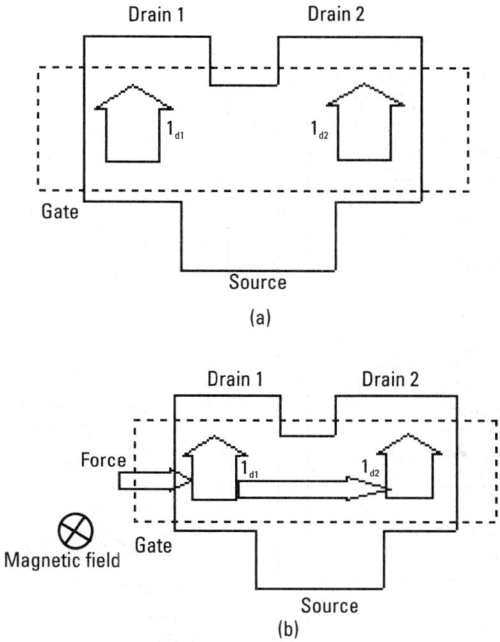

Figure 5.9 (a) Schematic of MAGFET. (b) Operation of MAGFET.

trokinetic methods that have been demonstrated in noninvasive biological cell identification, interrogation, and species separation applications.

5.5.1 Bioanalysis Platform

Keliman et al. have suggested a bioanalysis system that will be part of a future low-power bioanalysis platform [21]. The analysis technique exploits dielectrophoresis (DEP), an electrokinetic phenomenon that has demonstrated novel and noninvasive biological cell identification, interrogation, and species separation capabilities. Various electrode configurations have been developed and implemented, each of which can manipulate cells in a specific manner, and test microstructures have been built by fabricating the electrodes using a standard CMOS process.

A basic block diagram of such a proposed platform is shown Figure 5.10 [21]. The lexel array custom IP block is used to synthesize a specifically shaped electric field(s) for the manipulation of cells. This IP block is fabricated using a CMOS process, and operates in conjunction with an integrated microfluidic device. The function of this platform is as follows. A biological sample (cell or bacterial suspension) will be suitably acquired and injected using a micropump or based on capillary action made to flow over the lexel array within the microfluidic

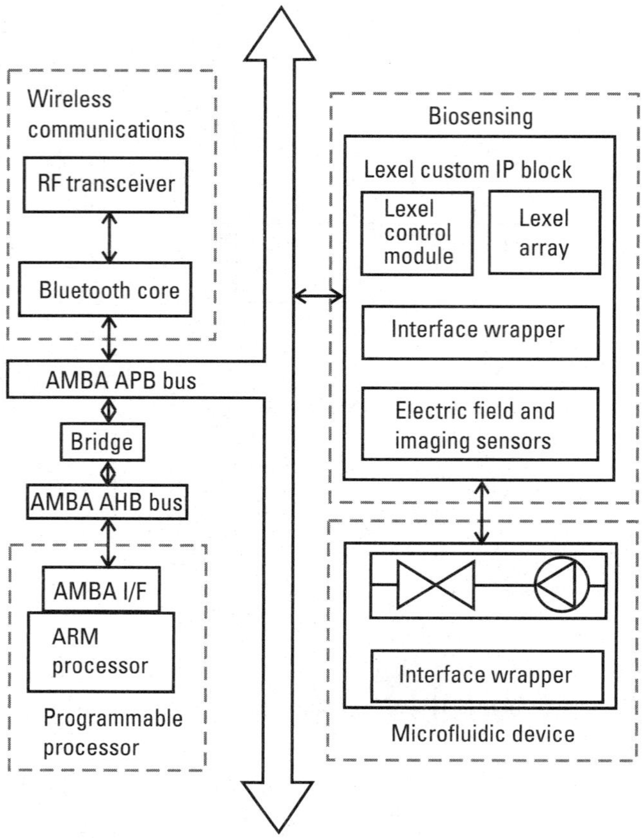

Figure 5.10 The bioanalysis platform. (*From:* [21]. Reprinted with permission.)

device. The nonuniform field(s) generated by the lexel array manipulates the cell sample in a specific fashion, in accordance with instructions from the ARM processor, and the resulting signals from the lexel control module. The electric field and image sensors collect data for feedback control, as well as for the purposes of identification and characterization of the sample by the ARM processor. A variety of data will be transmitted to a base station, where additional analysis of the biocell and data storage will occur.

The bioanalysis platform has the capability to implement a variety of electric fields from a single electrode structure in order to effect DEP manipulation of particles [21]. A planar array of microelectrodes with the ability to impose arbitrary voltages (potentials) to each individual electrode was established as the best method to realize this versatility of the electrode structure. In order to keep the number of control signals manageable, while still having the ability to impose arbitrary voltages (potentials) to each individual electrode, a conventional sample and hold circuit comprised of a transmission gate and capacitor is implemented at each lexel, where the lexel is the top plate of the capacitor. This programmable and reconfigurable electrode structure is referred to as the lexel array.

The lexel array is a checkerboard pattern of planar microelectrodes [21]. Figure 5.11 shows the arrangement of the electrodes in a sample section taken from the lexel array. The checkerboard pattern allows for the maximum electrode density per unit area due to the constraints of implementing the sample and hold circuit using MiM capacitors in the CMOSP18 fabrication process. The lexels in the CMOSP18 process are octagonal in shape to accommodate the routing conductors. The distance across the lexels and the spacing between the lexels is 10 μm. A ground plane covers the spaces between lexels to shield the particles from the stray fields of the active devices and routing conductors [21].

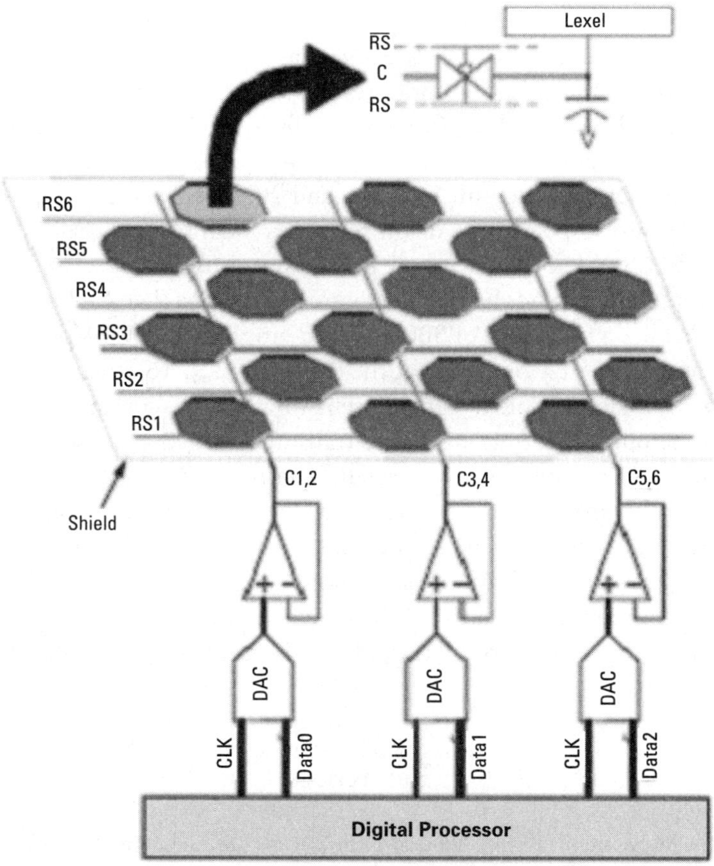

Figure 5.11 Block diagram showing the arrangement of the lexel array. (*From:* [21]. Reprinted with permission.)

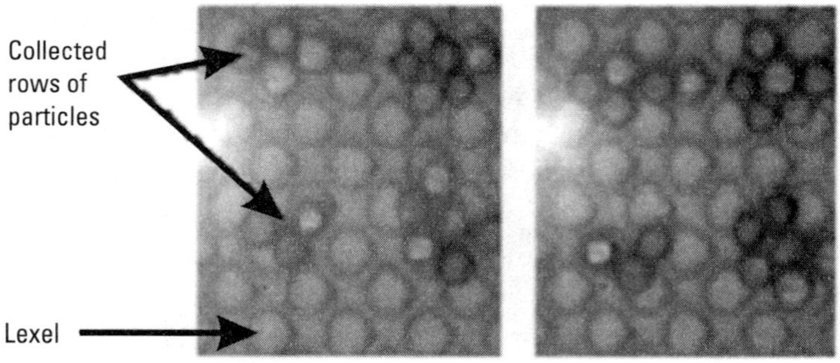

Figure 5.12 Collection and translation of a chain of particles using an interdigitated implementation. (*From:* [22]. Reprinted with permission.)

5.5.2 Experimental Tests

A fluidic chamber of approximately 10 μL has been created to contain the suspending medium and particles under test. The lexel array forms the bottom of the chamber and silicon rubber forms the walls of the chamber. A small piece of glass covers the chamber. Polystyrene microspheres with a relative permittivity of 2.5 and diameters of 9.6 pm, 14.6 pm, and 20.9 pm are used as test particles. The particles are suspended in deionized water with a measured conductivity between 1.3–1.9 pS/cm. A reflectance microscope is used to observe the particles from the top of the fluidic chamber. Tests are prepared so that the resulting signal on each lexel has a frequency of 300 kHz and amplitude of 3.3 Vpp. Each look-up table contains 64 entries to all of the columns [22]. To arrange the particles into rows, a single sinusoidal signal is sent to all of the columns. To optimize the FPGA resources that are used, the transmission gates in rows that receive the sinusoidal signal are always activated. The transmission gates in rows that are connected to ground are activated only when the sinusoidal signal is at 0V. The columns and rows of particles are conveyed by appropriately switching the columns and rows that receive the sinusoidal signal. Figure 5.12 shows the collection of a column of particles and the conveyance of the column to the right. Similar conveyance can also be achieved with particles arranged in rows that convey vertically.

References

[1] Talary, M. S., and R. Pethig, "Optical Technique for Measuring the Positive and Negative Dielectrophoretic Behavior of Cells and Colloidal Suspensions," *IEE Proceedings of Scientific Measurements and Technology*, Vol. 14, No. 5, September 1994, pp. 395–399.

[2] Burt, J. P. H., et al., "An Optical Dielectrophoresis Spectrometer for Low Frequency Measurements on Colloidal Suspensions," *Journal of Physics E (Scientific Instruments)*, Vol. 22, 1989, pp. 952–957.

[3] Price, J. A. R., J. P. H. Burt, and R. Pethig, "Applications of a New Optical Technique for Measuring the Dielectrophoretic Behavior of Microorganisms," *Biochimca et Biophysica Acta*, Vol. 964, 1988, pp. 221–230.

[4] Eyal, S., and S. R. Quake, "Velocity-Independent Microfluidic Flow Cytometry," *Electrophoresis*, Vol. 23, 2002, pp. 2653–2657.

[5] Johnson, G. E., B. P. Dorman, and F. H. Ruddle, "Ratio of Analog Pulse Height Module for Fluorescence Activated Cell Sorter," *Journal of Review of Scientific Instruments*, Vol. 50, No. 1, 1979, pp. 109–110.

[6] Fu, A. Y., et al., "A Microfabricated Fluorescence-Activated Cell Sorter," *Journal of Nature Biotechnology*, Vol. 17, 1999, pp. 1109–1111.

[7] Gawad, S., L. Schild, and P. Renaud, "Micromachined Impedance Spectroscopy Flow Cytometer for Cell Analysis and Particle Sizing," *Journal of Lab on a Chip*, Vol. 1, 2001, pp. 76–82.

[8] Fuller, K. C., et al., "Microfabricated Multi-Frequency Particle Impedance Characterization Systems," *Proceedings of In Micro Total Analysis Systems*, 2000.

[9] Sohn, L. L., et al., "Capacitance Cytometry: Measuring Biological Cells One by One," *Proceedings of the National Academy of Sciences USA*, Vol. 97, 2002, pp. 10687–10690.

[10] Ayliffe, H. E., A. B. Frazier, and R. D. Rabbitt, "Electric Impedance Spectroscopy Using Microchannels with Integrated Metal Electrodes," *IEEE Journal of Microelectromechal Systems*, Vol. 8, No. 1, 1999, pp. 50–57.

[11] Gawad, S., et al., "Fabrication of a Microfluidic Cell Analyzer in a Microchannel Using Impedance Spectroscopy," *1st Annual International Conference on Microtechnologies in Medicine and Biology*, 2000, pp. 297–301.

[12] Popovic, R., and H. Baltes, "A CMOS Magnetic Field Sensor," *IEEE Journal of Solid State Circuits*, Vol. 18, No. 4, 1983, pp. 426–428.

[13] Busatto, G., et al., "MAGFET Based Current Sensing for Power Integrated Circuit," *Journal of Microelectronics Reliability*, Vol. 43, 2003, pp. 577–583.

[14] Liu, S., J. Wei, and G. Sung, "SPICE Macro Model for MAGFET and Its Applications," *IEEE Trans. on Circuits and Systems—II: Analog and Digital Signal Processing*, Vol. 46, No. 4, 1999.

[15] Medoro, G., et al., "A Lab-on-a-Chip for Cell Detection and Manipulation," *Proceedings of IEEE Sensors*, Vol. 1, Orlando, FL, 2002, pp. 472–475.

[16] Romani, A., et al., "Capacitive Sensor Array for Localization of Bioparticles in CMOS Lab-on-a-Chip," *IEEE International Solid-State Circuits Conference (ISSCC 04)*, 2004, pp. 456–457.

[17] Medoro, G., et al., "CMOS-Only Sensors and Manipulation for Microorganisms," *Proceedings of IEDM*, 2000, pp. 415–418.

[18] Manaresi, N., et al., "A CMOC Chip for Individual Manipulation and Detection," *IEEE International Solid-State Circuits Conference (ISSCC 03)*, 2003, pp. 486–488.

[19] Medoro, G., et al., "A Lab-on-a-Chip for Cell Detection and Manipulation," *IEEE Sensors Journal*, Vol. 3, No. 3, 2003, pp. 317–325.

[20] Jones, T. B., *Electromechanics of Particles*, Cambridge, U.K.: Cambridge University Press, 1995.

[21] Keilmnt, J. R., G. A. Jullient, and K. Kaler, "A SoC Bio-Analysis Platform for Real-Time Biological Cell Analysis-on-a-Chip," *IEEE International Workshop on Biomedical Circuits & Systems*, Calgary, AB, Canada, 2003.

[22] Keilmnt, J. R., G. A. Jullient, and K. Kaler, "A Programmable ac Electrokinetic Micro-Particle Analysis System," *IEEE International Workshop on Biomedical Circuits & Systems,* Banff, AB, Canada, 2004, pp. S2.3-1–S2.3-12.

6

CMOS-Based Lab-on-a-Chip

Lab-on-a-chip is a new technology that is changing the traditional way by which biological samples are inspected in laboratories during analyses. Lab-on-a-chip includes miniaturized and smaller integrated systems. Miniaturization allows for replacement of traditional bulky and expensive equipment with cheaper and portable equipment. In cell biology, the use of microfabricated structures employing active substrates and integrated sensors [1] is still at the experimental stage. However, these techniques have already been tested for identification of objects at the microscopic level so as to envision several applications in the replacement of traditional optical inspection techniques. The presence of a silicon substrate and the complementary metal-oxide semiconductor (CMOS) technology allows the functional integration of sensors, signal conditioning, and processing circuits and the development of a fully electronic, integrated lab-on-a-chip. This chapter covers and presents the state of the art in the area of CMOS-based lab-on-a-chip for biological applications.

6.1 PCB Lab-on-a-Chip for Micro-Organism Detection and Characterization

In 2002 Medoro et al. [2, 3] proposed the first lab-on-a-chip approach for electronic manipulation and detection of micro-organisms. The approach combines dielectrophoresis with impedance measurement to trap and move particles while monitoring their location and quantity in the device. The prototype has been realized using standard printed circuit board (PCB) technology. The actuation part in this approach is based on the DEP moving cage [2]; the sensing can be performed using a transimpedance amplifier. Figure 6.1 shows a picture of the complete lab-on-a-chip system [2]. It consists of three main parts: a laptop (PC), a motherboard, and an actuator/sensor device. The sensor/actuator is plugged into a motherboard, which controls and generates the proper phased voltages for each electrode in the device in order to create and control the DEP cages and to enable particle sensing. The actuation and sensing operations have been controlled and automated by a software tool running on a microcomputer (PC). The microchamber (Figure 6.2) is comprised of a top plate and a printed circuit board (PCB) on the bottom for support. The top plate is both conductive and transparent; it serves as an electrode and it is electrically connected to the PCB device by means of a conductive glue. A spacer (realized by two optical fibers) determines the chamber height, while a silicon elastomer gasket delimits and seals the microchamber on the sides [2, 4].

6.2 Actuation

The DEP cage in the spatial region above an electrode can be obtained by connecting the associated electrode and the microchamber lid to an out-of-phase

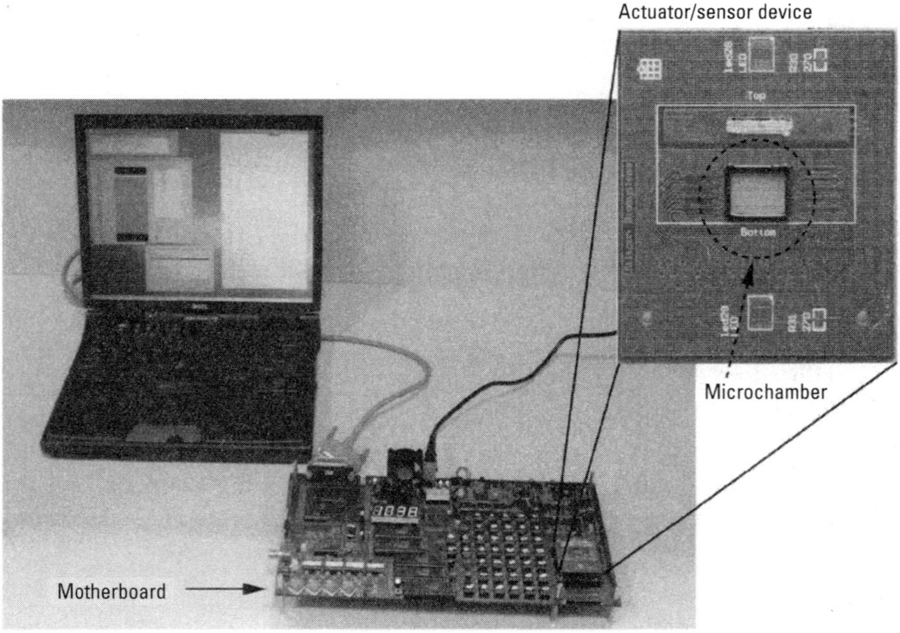

Figure 6.1 System setup: the actuation and sensing phases are executed under software control by means of a PC. A motherboard is used to generate and apply the stimuli to each electrode in the device. (*From:* [1]. Reprinted with permission.)

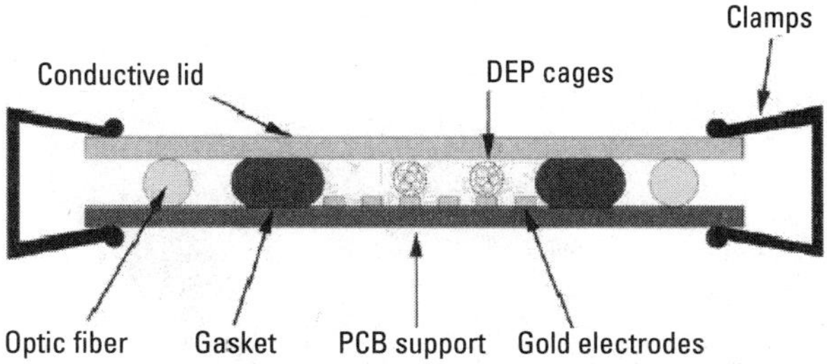

Figure 6.2 Cross-section of the device. A conductive lid and a PCB support close the microchamber on the top and on the bottom; two optic fibers are used as a spacer and a gasket is used to delimit the microchamber on the sides. (*From:* [1]. Reprinted with permission.)

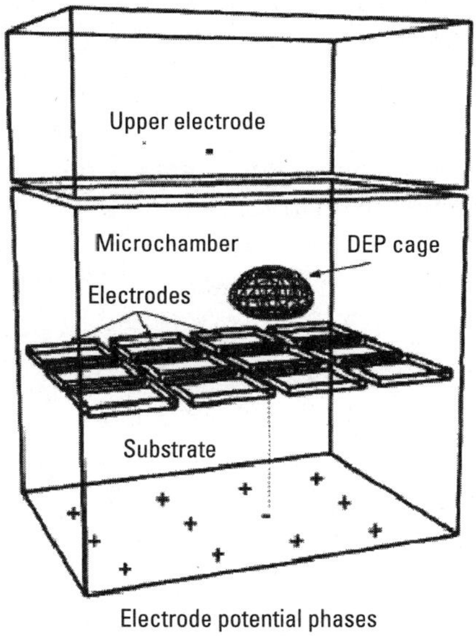

Figure 6.3 The approach for establishing a DEP cage. (*From:* [1]. Reprinted with permission.)

(180° phase) sinusoidal voltage, while the neighboring electrodes are connected to an in-phase sinusoidal voltage. A field minimum is then obtained in the liquid corresponding to a DEP cage in which one or more particles are trapped and levitated. By changing the applied pattern voltage to the electrodes, the DEP cage can be independently moved around the device plan [5]. Under software control, a cylinder-shaped DEP cage can be realized above each generic electrode.

The prototype device has 39 electrodes to realize up to 19 DEP cages. By sequentially changing the electrode activation pattern, each DEP cage can also be independently moved from one electrode to the next along the whole microchamber while dragging the trapped particles (Figure 6.4). After the electrode

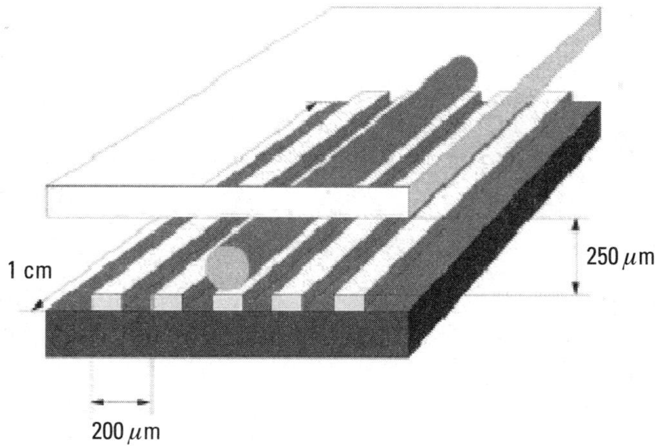

Figure 6.4 Schematic view of the prototype device, where a cylinder-shaped DEP cage corresponding to the central electrode. (*From:* [1]. Reprinted with permission.)

activation pattern has been changed, the cage disappears from the original location and reappears, for example, in correspondence to one of the neighboring electrodes. At the beginning the particles remain in the original location due to inertial and viscous forces. Since they are still in the attraction basin of the shifted cage, they will end up toward its center. For spherical geometries, the first order equation time-averaged DEP force $\left\langle \vec{F}_{DEP} \right\rangle$ is defined as follows:

$$\left\langle \vec{F}_{DEP} \right\rangle = 2\pi a^3 \varepsilon_m \operatorname{Re}[\underline{K}] \nabla E_{rms}^2 \tag{6.1}$$

where

$$\underline{K} = \left[\frac{\varepsilon_p^* - \varepsilon_m^*}{\varepsilon_p^* + 2\varepsilon_m^*} \right] \tag{6.2}$$

$\underline{K}$ is the complex Clausius-Mossotti factor, and ε_p^* and ε_m^* are defined by (6.3) and (6.4), respectively:

$$\varepsilon_p^* = \varepsilon_p - j\frac{\sigma_p}{\omega} \tag{6.3}$$

$$\varepsilon_m^* = \varepsilon_m - j\frac{\sigma_m}{\omega} \tag{6.4}$$

where σ_p and σ_m are the conductivity of the particle and the medium, respectively, and ω is the angular frequency of the applied electric field (refer to Chapter 2 for complete details about the dielectrophoresis force).

For negative DEP (i.e., $\underline{K}$ is negative), $\underline{K}$ can be approximated at a low frequency by:

$$\underline{K} = \left(\frac{\sigma_p - \sigma_m}{\sigma_p + 2\sigma_m} \right) \tag{6.5}$$

while at a high frequency, $\underline{K}$ is given as:

$$\underline{K} = \left(\frac{\varepsilon_p - \varepsilon_m}{\varepsilon_p + 2\varepsilon_m} \right) \tag{6.6}$$

Thus, particle levitation is possible at low frequencies if $\sigma_p < \sigma_m$, and at high frequencies if $\varepsilon_p < \varepsilon_m$.

The viscosity force (F_{Vis}) is defined as:

$$F_{Vis} = A\eta \frac{dv}{dy} \tag{6.7}$$

where A is the particle's area (for a spherical particle $A = \pi a^2$), η is the viscosity coefficient, and $\frac{dv}{dy}$ is the viscosity gradient [6].

In fact, the dynamic of the particle is strictly influenced by its size. From (6.1) and (6.7), the dielectrophoretic force is proportional to the particle volume ($F_{Dep} \alpha a^3$), while the viscous force is proportional to the particle surface ($F_{Vis} \alpha a^2$). The ratio between dielectrophoresis and the viscous force is therefore proportional to $\frac{F_{Dep}}{F_{Vis}} \alpha$. Accordingly, the timing associated with the electrode programming sequence has to be chosen considering the slower particle dynamics, as scaling down the particle geometry causes the dynamics to slow down. After a cage is shifted, one needs to wait for the slowest particles to reach the new trapping point (i.e., for 3-μm diameter particles, 5–10 seconds are needed, while for 50-μm diameter particles, 2 seconds are enough) [1]. Otherwise, if the cage shift is too fast, some particles may be left behind.

6.3 Impedance Sensing

The sensing part in this approach can be performed for any electrode by multiplexing it between providing the electrical stimulus and as the input to a transimpedance amplifier, while all the other electrodes are connected to ground and a sinusoidal stimulus is applied to the lid (Figure 6.5). During the sensing task the actuation is switched off. However, levitation is preserved since the sensing operation is much faster than the time required by the particles to fall down to the chamber floor (about 1 second is needed for 3-μm microbeads). The presence of the particle above the electrodes induces a perturbation of the electric field that can be detected by impedance sensing [7, 8].

The transfer function of the sensing circuit is described by:

$$\frac{v_{out}}{v_{in}} = -\frac{R_F}{R_M}\left(\frac{1 + j\omega R_M C_M}{1 + j\omega R_F C_F}\right) \tag{6.8}$$

where v_{out} is the circuit output voltage, v_{in} is the input stimulus, R_F and C_F and are the feedback resistance and capacitance, and R_M and C_M are the resistance and capacitance between the lid and the sensing electrode. In the low-frequency range, the response is dominated by the resistive effects and (6.3) can be approximated by

$$\frac{v_{out}}{v_{in}} = -\frac{R_F}{R_M} \tag{6.9}$$

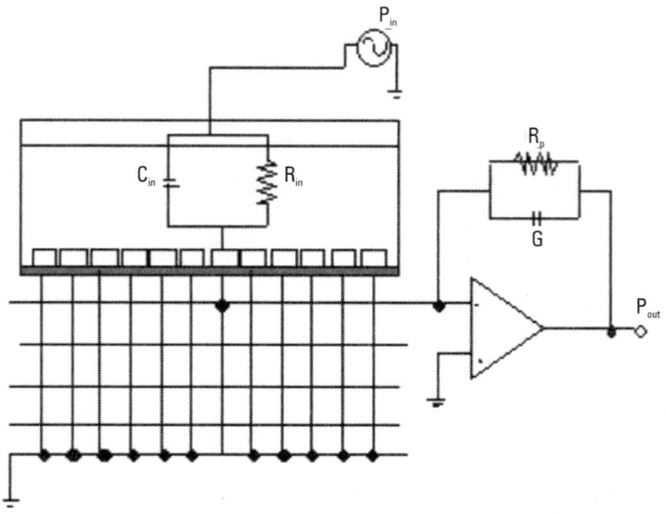

Figure 6.5 Impedance sensing. (*From:* [8]. Reprinted with permission.)

In the high-frequency range, the response is dominated by the capacitive effects and (6.8) can be approximated by

$$\frac{v_{out}}{v_{in}} = -\frac{C_M}{C_F} \tag{6.10}$$

From (6.5) and at low-frequency stimuli, a particle can be trapped if its conductivity is lower than the medium ($\sigma_p < \sigma_m$). Accordingly, the effective conductivity between the sensing electrode and the lid decreases. Therefore, if we use the same frequency for actuation and sensing, R_M increases with the presence of trapped particles

$$R_M\left(\text{with particle}\right) > R_M\left(\text{without particle}\right) \tag{6.11}$$

From (6.9) and (6.11), we can predict that the voltage amplitude of the transimpedance amplifier output decreases when particles are trapped above the electrode. Similarly, at the high frequency range, from (6.6), a particle can be trapped if its permittivity is lower than the medium ($\varepsilon_p < \varepsilon_m$), and thus

$$C_M\left(\text{with particle}\right) < C_M\left(\text{without particle}\right) \tag{6.12}$$

which substituted in (6.10) yields a lower output voltage amplitude.

The advantages of this lab-on-a-chip can be summarized as follows:

1. It is possible to build a map of particle distribution by repeating the sensing operation for each electrode in the device.
2. The volume occupied by the particles can be measured by measuring the output variation. Hence, an estimation of the quantity of trapped particles for each cage may be possible.

However for this purpose, a precalibration procedure is required to associate a volume quantity to the sensing value. Thus, if the cell specie in the microchamber is known and previously characterized, cell quantification for each cage is possible. The concentration capabilities of the device can boost the sensitivity of the detection, which is useful when low concentration samples have to be analyzed. Sometimes without the recalibration, the sensing measure can give important information even if the micro-organism specie is not known. For example, contaminated water can be checked for micro-organisms. In this case, one has to assess whether micro-organisms are present in the sample or not. This could be revealed by decreasing sensing output after merging DEP cages.

Figure 6.6 shows an experiment demonstrating that the PCB-based lab-on-a-chip is suitable for directing single microspheres to a single identified target cell [9]. The figure shows three cationic microspheres (M1, M2, and M3) and two K562 cells entrapped in five independent spherical DEP cages. Only one of the two cells was the three microspheres' cellular target. The device first moved microsphere M1, generating the cell-microsphere complex shown in Figure 6.6(d); then it moved microsphere M2, generating the K562-M1M2 complex shown in Figure 6.6(e). Finally, it moved microsphere M3 for a further targeting of the K562 cell, obtaining the K562-M1M2M3 complex shown in Figure 6.6(f). The buffer employed was 280-milliMolar (mM) mannitol and 6.5-mM potassium chloride. The experimental conditions were 50 kHz for the AC electric field used to create dielectrophoretic patterns, and 37°C for the supernatant. The passivation layer provides protection against metal contamination and inhibits electrolysis.

6.4 CMOS Lab-on-a-Chip for Micro-Organism Detection and Manipulation

Manaresi et al. [10, 11] proposed another CMOS lab-on-a-chip for cell manipulation and detection based on standard 0.35-μm CMOS technology. This lab-on-a-chip microsystem consists of three main units: the actuation unit, which

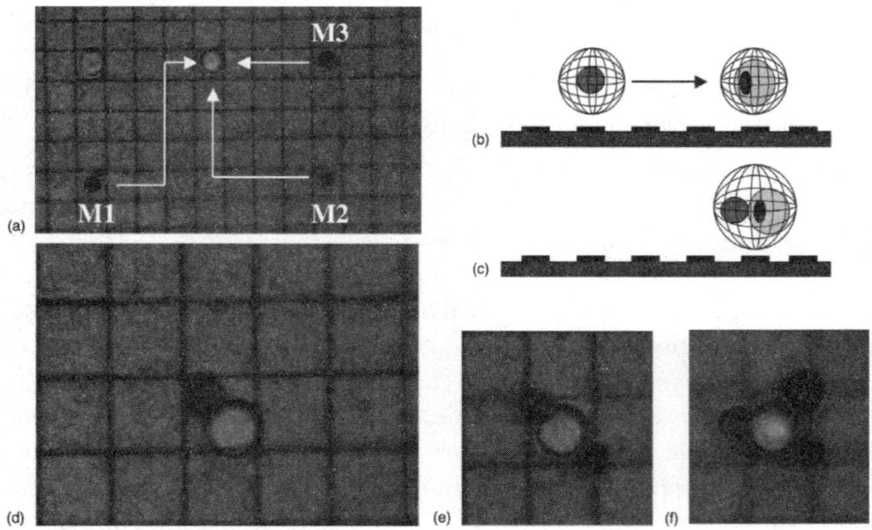

Figure 6.6 (a–c) Programmed sequential interactions between three cationic microspheres (M1, M2, and M3) and one target K562 cell. Moving M1 produced (d) the K562-M1 complex. Moving M2 and M3 produced the (e) K562-M1M2 and (f) K562-M1M2M3 complexes. (*From:* [9]. Reprinted with permission.)

creates the DEP cage; the sensing unit; and a microfludic chamber. The lab-on-a-chip architecture (Figure 6.7) is based on a two-dimensional array of microsites, the purpose of which is to:

1. Generate the electric field necessary to create dielectrophoretic cages.
2. Detect the presence of single particles or clusters trapped in cages by using optical sensing.

Each microsite consists of an actuation electrode, implemented with a top metal plate, and underlying embedded circuitry for programming and detection. Microsites can be addressed in a random access mode by means of row and column decoders for both actuation and sensing.

The presence of particles is detected by photodiodes embedded in the substrate that measure signal variations from uniform light impinging on the chip surface. The mode of operation consists of three phases: programming,

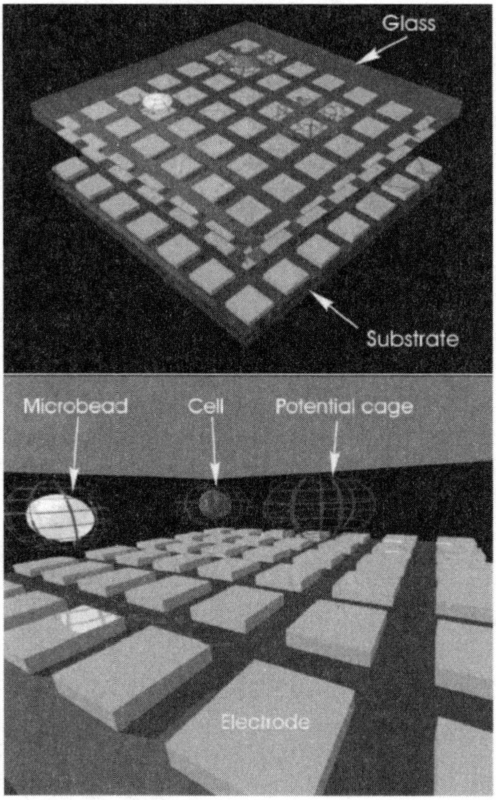

Figure 6.7 The lab-on-a-chip architecture for cell manipulation and sensing. (*From:* [10]. Reprinted with permission.)

actuation, and sensing. During programming the actuation pattern is stored in the microsites of the array to determine DEP cage number, profile, and shape. In the actuation phase each electrode is energized by either an in-phase or counter phase sinusoidal actuation voltage signal, according to the programmed patterns. During the sensing phase, the optical image of the array is grabbed. The general architecture of the chip is sketched in Figure 6.7. The chip contains the following blocks:

- An array of microsites, composed of 320 elements of 20-micrometer pitch;
- 9-bit static column/row decoders for random access;
- A bias generator block;
- A readout circuit block.

Referring to Figure 6.8, the row and column circuits provide the logic signals to program and read out each microsite [11].

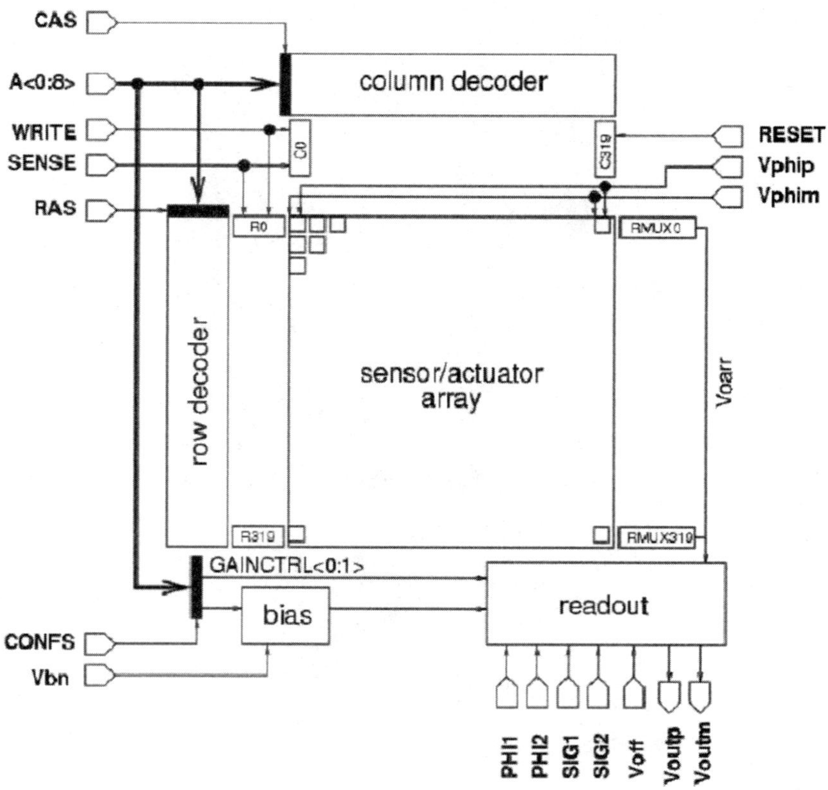

Figure 6.8 Chip architecture. (*From:* [11]. Reprinted with permission.)

The actuation part in this lab-on-a-chip is based on the same DEP cage technique [1], as mentioned earlier in Section 6.1. The sensing junction depends on those particles in the sample that can be detected by the changes in optical radiation impinging on the photodiode associated with each microsite. During the sensing, the actuation voltages are halted to avoid coupling with the pixel readout. However, due to inertia, the cells keep their position in the liquid. Thus, actuation and sensing phases are always kept nonoverlapping. On the contrary, programming and actuation may be concurrent: actuation patterns can be changed in real time while electrodes keep energizing the system [11]. A schematic diagram for a single microsite is shown in Figure 6.9 [11]. When the site is addressed and WRITE is activated, the metal 3 electrode can be switched between V_{phip} and V_{phim} by programming the memory element addressed by ROWW and COLW. The sensing circuit is an active-pixel optical sensor implemented with a $2 \times 17\ \mu m^2$ well-junction photodiode technique. The sensor array is read row-wise. After activating the sense junction, a pixel is addressed through ROWS and COLS, and the voltage, after integration, is sampled by the readout circuit through V_{orow}. While the pixel is still addressed, RESET goes high, and the reset voltage is sampled also by the readout circuit. The difference between the two voltages is amplified and outputted as a differential voltage $V_{outp} - V_{outm}$. The subtraction of the reset voltage allows one to compensate the pixel's fixed pattern noise.

Figure 6.10 shows the chip in the microfluidic package [11, 12]. After a conductive-glass lid with SU8 walls is glued to the chip to define the microchamber, the die is mounted directly on the printed circuit board (PCB) with the chip-on-board technique [13, 14]. The lid is spaced about 85 μm from the chip surface; lid height is not critical. The strength of the DEP cage mostly depends on its height above the array; since the cage is near the array, the field gradients are stronger (the lower cage results in a stronger the force).

In turn, cage height increases with lid height and decreases with lid-voltage amplitude. Simulation results show that increasing lid height from 50 to 80 μm while increasing lid-voltage amplitude from 3.3 to 6.6V, results in cage height being approximately the same (i.e., about 15 μm) [12, 13]. Considering a cell radius of about 10 μm, this is somewhat of a lower bound in order to prevent cell contact with the array. The interdevice variations of microchamber height (e.g., ±10% due to SU8 thickness variations), or intradevice nonplanarity of the lid, have a negligible impact on DEP cage strength, such as for constant lid-voltage amplitude; DEP force variation with microchamber height is low. For example, keeping a 3.3-V lid voltage, when increasing the lid height as above from 50 to 80 μm (i.e., 60%), the horizontal DEP force decreases due to a variation in cage height of less than −50%. Epoxy resin is used for bonding wire protection [13]. The fluidic inlet is provided by a capillary connected to an opening in the microchamber lateral wall and sealed with a drop of insulating glue. A drop of

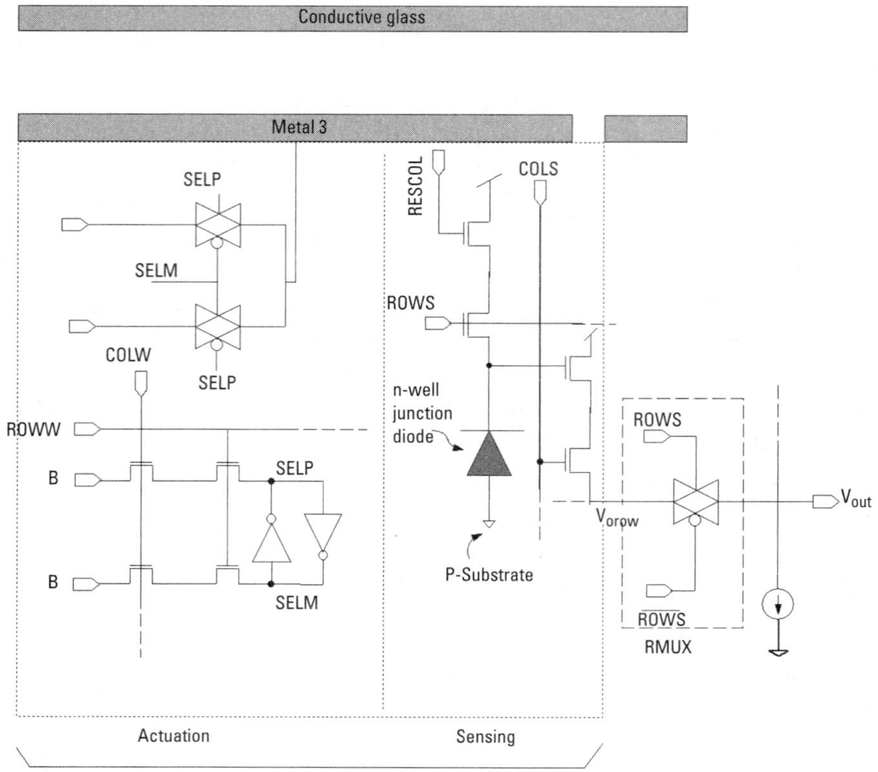

Figure 6.9 Schematic of a microsite. (*From:* [11]. Reprinted with permission.)

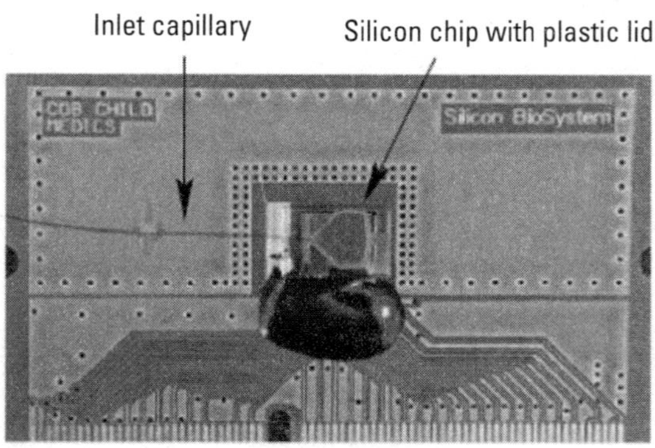

Figure 6.10 The lab-on-a-chip. (*From:* [11]. Reprinted with permission.)

conductive glue is used to electrically connect the lid electrode to the PCB. The die photograph with the microfluidic chamber is shown in Figure 6.11. Figure 6.12 shows the corresponding image acquired with the embedded sensors after

Figure 6.11 CMOS biosensor/actuator for cell analysis. The fluidic microchamber packaging is implemented by double bonding the ito-coated glass, patterned with dry-resist film, to a CMOS chip. (*From:* [10]. Reprinted with permission.)

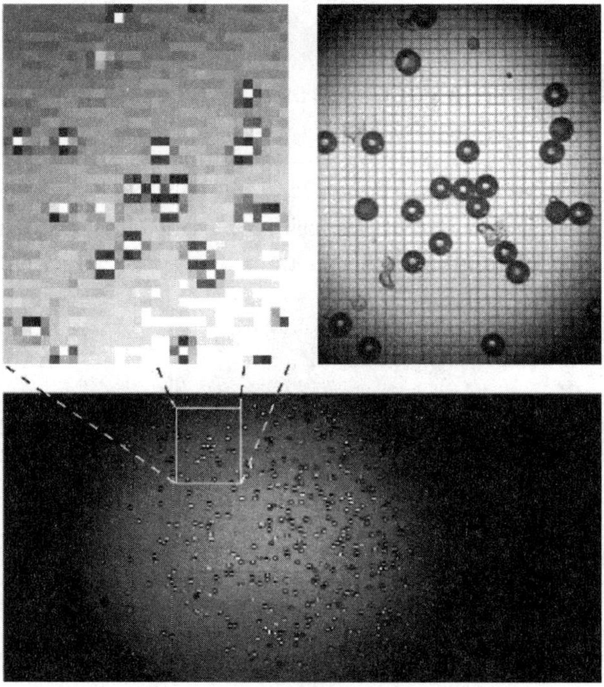

Figure 6.12 Comparison of embedded sensor (top left) and optical microscope (top right) images of 50-μm polystyrene beads. (*From:* [10]. Reprinted with permission.)

a sample (50-μm polystyrene beads in water) has been flushed into the microchamber; the beads are randomly distributed. DEP cages are activated with 3.3V.

Three snapshots of the selective motion of one bead are reported in Figure 6.13. The top line shows the images acquired by the microscope while the bottom line displays the corresponding programming pattern, where gray and white squares indicate the electrodes receiving V_{phim} and V_{phip}, respectively; the lid is always proportional to V_{phim}. The time for particles to complete one step is approximately 2 seconds.

Figure 6.14 shows how, after applying a pattern implementing an array of DEP cages, the microbeads are arranged correspondingly, and how they can be detected with the embedded optical sensors. Figure 6.15 shows how the chip can control two K562 tumor cells, as it shows the mating and separation of two K562 tumor cells. Two cages trapping the selected cells [Figure 6.15(a)] are first merged into a three-electrode cage [Figure 6.15(b)], which is then shrunk to a single electrode cage [Figure 6.15(c, d)], forcing the two cells to contact. The cage is then enlarged [Figure 6.15(e)] and the cells lose contact until they are finally separated again in two distinct cages [Figure 6.15(f)] [10, 11]. The chip also was used to characterize a K562 human erythro-leukemia cells and Yarrowia lypolytica yeasts, respectively. The results are shown in Figure 6.16 [13].

6.5 CMOS Lab-on-a-Chip for Neuronal Activity Detection

Neurons are biological cells specialized in transmission and processing information [15]. The elementary neural signals are action potentials, which are transient changes of the voltage drop across the cell membrane with a typical shape. The related peak-to-peak amplitudes of this transmembrane voltage approximately

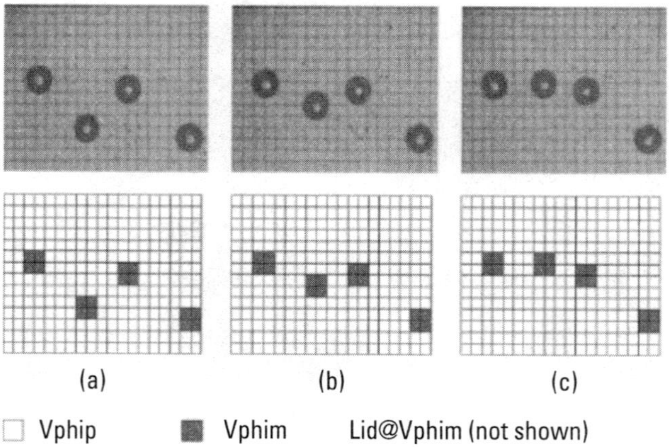

Figure 6.13 (a–c) Individual manipulation of a 50-μm polystyrene bead in water; 3.3V at 800-kHz phases. (*From:* [10]. Reprinted with permission.)

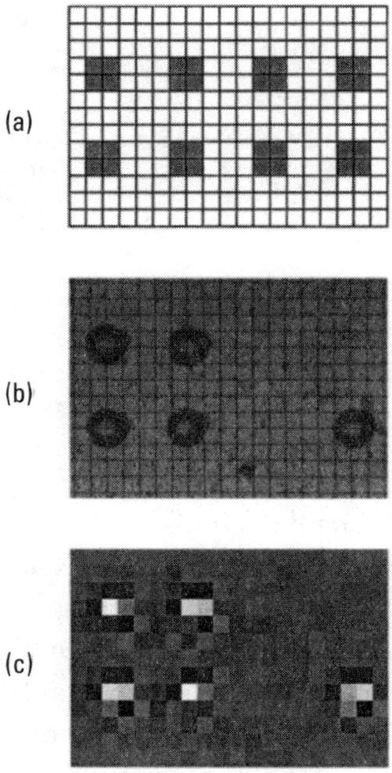

Figure 6.14 Manipulation and detection of 50-μm polystyrene beads. (a) Actuation pattern, (b) microscope image, and (c) embedded optical sensors image. (*From:* [10]. Reprinted with permission.)

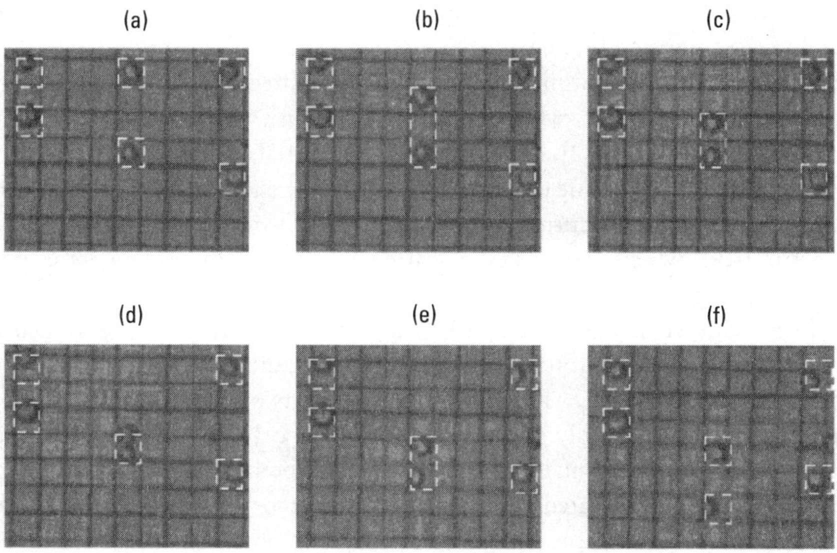

Figure 6.15 (a–c) Mating and (d–f) separation of K562 tumor cells in 280-mM mannitol in water, 3.3 Vpp at 500-kHz phases. Cage electrodes are enclosed by dashed lines. (*From:* [9]. Reprinted with permission.)

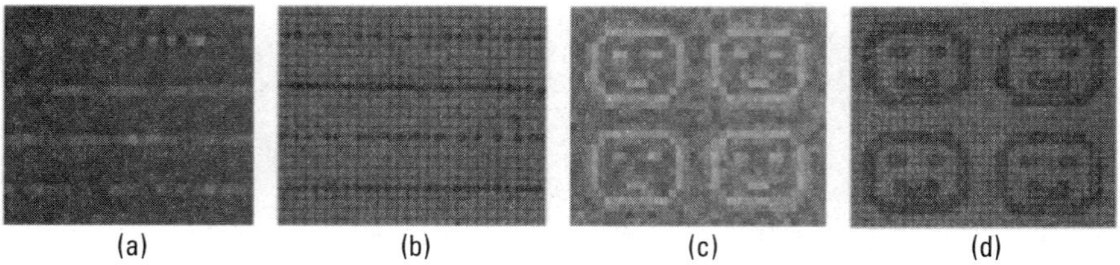

Figure 6.16 An optical sensor image of dielectrophoretically trapped K562 human erythro-leukemia cells (a) is compared with a corresponding microscope image of the sensor array (b). In (c), an image of trapped Yarrowia lypolytica yeasts obtained with capacitive sensors matches the microscope image (d). (*From:* [13]. Reprinted with permission.)

amount to 70 mV. Today's standard tools used to characterize this parameter are patch pipettes [16] or microelectrodes [17]. These tools have led to great achievements in the field of neural cell monitoring. However, these techniques suffer from drawbacks, such as that they require an elaborate mechanical setup, which allows the monitoring of very few cells in parallel. Thus, they are generally not suitable to fulfill high-throughput requirements. Also, long-term recording is restricted due to the invasive type of contact, reducing the lifetime of the cell. Extracellular, and therefore, noninvasive, recording techniques have opened a way to circumvent these drawbacks [18–20]. The noninvasive recording is based on the fact that the action potentials to be considered correspond to sodium and potassium ion currents through ion channels in the cell membrane [15]. Consequently, the idea behind these approaches is to monitor these ion currents instead of the transmembrane voltage.

For extracellular recordings, cells are cultured directly on top of a transducing element (Figure 6.17), which is generally either a metallic electrode or an open-gate transistor [21, 22]. When an action potential occurs in a cell, the local flow of ions in and out of the cell causes the membrane to become nonuniformly polarized [23]. The ionic current flows across the cleft resistance and establishes an electric field, which induces electrical charge in the underlying transducer (the recorded signal).

Eversmann et al. proposed a CMOS lab-on-a-chip for extracellular recordings and detection of neuronal activity [18, 19]. Figure 6.18 shows the system setup and chip architecture. The sensor array consists of 128 × 128 pixels. The column decoder periodically selects one of the 128 columns of the pixel array. It also provides control signals, which determine whether the pixels within the selected column are operated in the readout mode or in a calibration mode. Calibration of all pixels is performed before they are operated in readout mode, and periodically repeated after a number of readout frames (typically, after 100 frames). The full frame readout rate is 2 kiloframes per second (2 kfps). Each of the 128 rows is connected to a separate readout amplifier. The outputs of

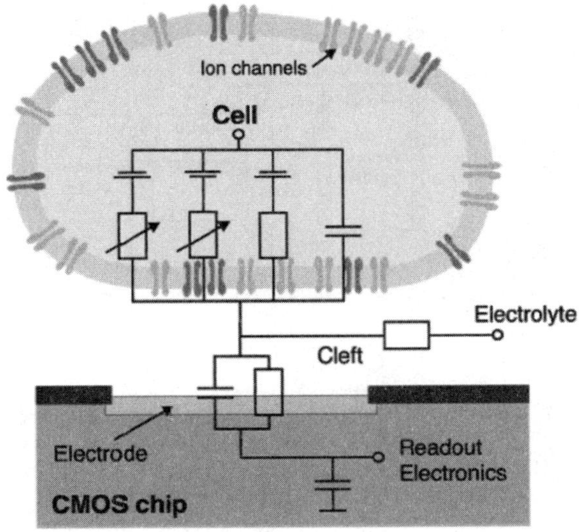

Figure 6.17 Schematic representation of the extracellular recording technique. During an action potential, ions flow across the cell membrane and cause a voltage change in the cleft, which can be measured using an extracellular electrode. The equivalent circuit shows the capacitance and ion conductances of the cell membrane, the electrode–electrolyte impedance, and the cleft resistance. (*From:* [20]. Reprinted with permission.)

these amplifiers are connected to the chip's output drivers or to dummy loads via 16 8-to-1 multiplexers. The 16 output drivers force an output current into current to voltage (I/V) converters arranged on an off-chip printed circuit board (PCB). Finally, the buffered output voltages of the I/V converters are converted by 16 analog-to-digital converters (ADCs) with an effective resolution above 8 bit; these ADCs are part of a PC-based measurement system. The resulting data rate amounts to 32 MS/s. A user interface program sorts and represents the acquired data. Moreover, this interface allows control of the timing of the applied control signals, the number of frames per calibration, and the power-up and power-down routines of the sensor array, by generating the corresponding control patterns. The latter are provided to the chip via magneto couplers, which drive the chip's logic interface. In Eversmann's lab-on-a-chip, a resistive temperature sensor is realized to control and regulate precisely the environment temperature, which influences biological processes such as neural activity; also, a regulation loop controls the chip's temperature utilizing a peltier element located underneath the sensor chip. In order to control the voltage of the nutrition electrolyte, an off-chip potentiostat circuit is operated in the measurement system (see Figure 6.18).

The extracellular recording capabilities are tested with neurons from a pond snail (*lymnaea stagnalis*). Before cultivation, the chips are cleaned, sterilized, and

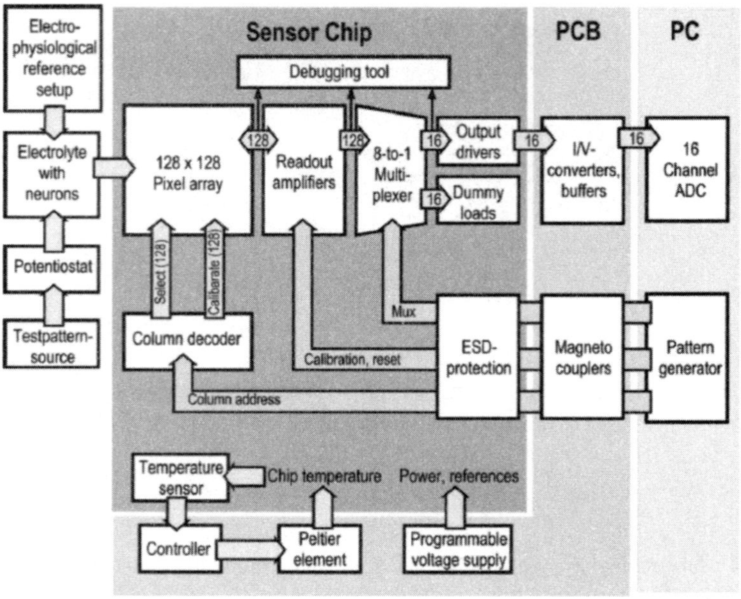

Figure 6.18 Architecture of a sensor array with a complete system. (*From:* [18]. Reprinted with permission.)

coated with a cell-adhesive layer of poly-l-lysine. Then, the neurons are extracted from the animals, placed on the sensor chip, and cultured for one day, as shown in Figure 6.19. Figure 6.20 shows the result of a neuron tapped by a microelectrode and stimulated by the injection of a constant current of 0.1 nA forced into the cell for 500 ms. For comparison purpose, the intracellular voltage is recorded with both the same trapped microelectrode and with the extracellular sensors. Bursts of four action potentials with amplitude of 60 mV are obtained.

In 2006, Heer et al. presented a new CMOS lab on a chip for bidirectional communication (stimulation and recording) [20]. Figure 6.21 shows a microphotograph of the monolithic CMOS lab on a chip. The 6.5-mm × 6.5-mm chip comprises 128 stimulation and recording-capable electrodes in an 8 × 16 array, and an integrated reference electrode [20]. The system approach is structured in a modular design as shown in Figure 6.22. Each pixel of the electrode array incorporates the signal-transducing electrode, a fully differential bandpass filter for immediate signal conditioning, selection switches to select either stimulation or recording mode, and a buffer for stimulation. The pitch of the pixel unit is 250 μm. For this lab-on-a-chip, the electrodes are realized during post-processing. Thus, the electrode pitch, size, and shape are very flexible and the electrode material can be selected from a large variety.

A digital control unit is integrated on chip. It controls the multiplexing, the electrode selection for stimulation, and the reset of single electrodes, and it

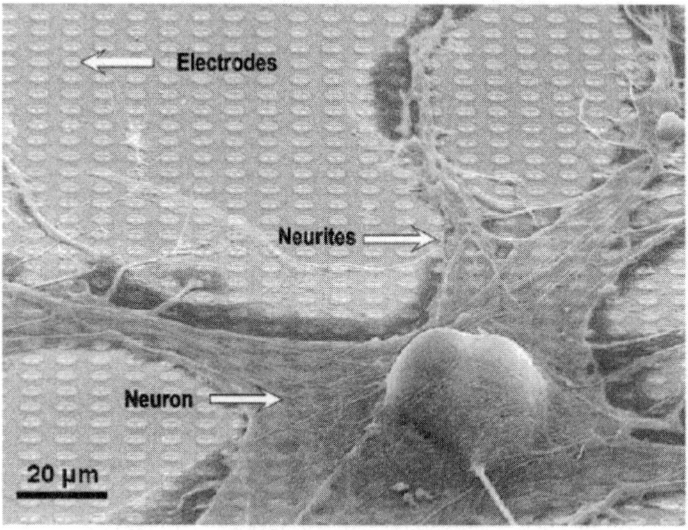

Figure 6.19 Snail neuron on a sensor chip in culture. (*From:* [18]. Reprinted with permission.)

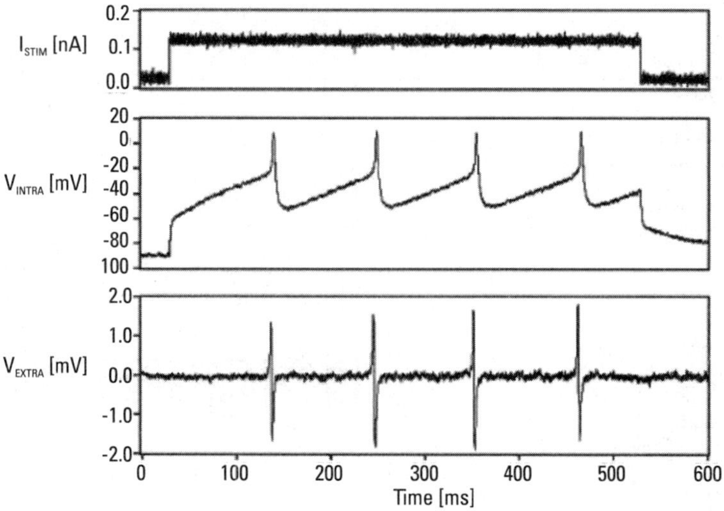

Figure 6.20 Measured data from a snail neuron. The neuron is stimulated and monitored for reference purposes with a microelectrode. Stimulation current I_{STIM}, intracellular potential V_{INTRA}, and extracellular potential V_{EXTRA} recorded with sensor array. (*From:* [18]. Reprinted with permission.)

contains the successive-approximation registers of the analog-to-digital (A/D) converters and the interfaces to the outside world. In comparison to Eversmann's

Figure 6.21 Micrograph of the chip. (*From:* [20]. Reprinted with permission.)

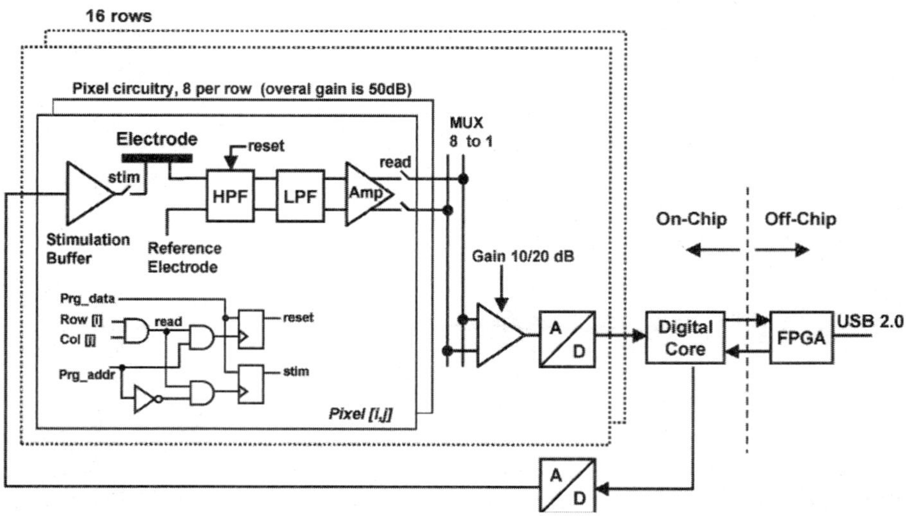

Figure 6.22 Block schematic of the overall system architecture. (*From:* [20]. Reprinted with permission.)

CMOS lab-on-a-chip [18, 19], this lab-on-a-chip, which implements filters and buffers at each electrode, offers important advantages:

1. The signal is amplified and filtered in close proximity of the electrodes, which makes the design less sensitive to noise and distortion picked up along connection lines.
2. A buffer renders the stimulation signal independent of the number of activated electrodes.
3. The highpass filter removes offset and drift of the biochemical signals and, therefore, the signal can be amplified before it is multiplexed.
4. The lowpass filter limits the noise bandwidth and works as an antialiasing filter for subsequent AD conversion.

The electrodes are continuously read out at a sampling rate of 20 kHz per electrode. An amplification of 1,000 or 3,000 can be selected. The overall power consumption of the chip is 120 mW at a 5-V supply, 20 mW of which are dissipated within the array. To control the environment's temperature, an on-chip temperature sensor, which monitors the chip operating temperature, is used. Also, to reduce the effect of the leakage currents, a full differential of the in-pixel readout circuitry is utilized. Furthermore, electromagnetic interference is also generally reduced in a fully differential architecture. The electrode units also provide stimulation capabilities. The circuitry at each electrode can be individually reset to its operating point in order to suppress artifacts evoked by the stimulation pulses from the stimulated electrode itself or from neighboring electrodes. The chip was then mounted and wire-bonded on a custom-designed printed circuit board (PCB) (see Figure 6.23). A water-resistant medical epoxy (EPOTEK 302-3M) was used to encapsulate the bond wires and pads. A glass ring forms a bath that contains a suitable amount of cell medium. The Heer's CMOS lab-on-a-chip is tested successfully with spontaneously firing chicken neurons. Figure 6.24 shows the extracellular recorded signal.

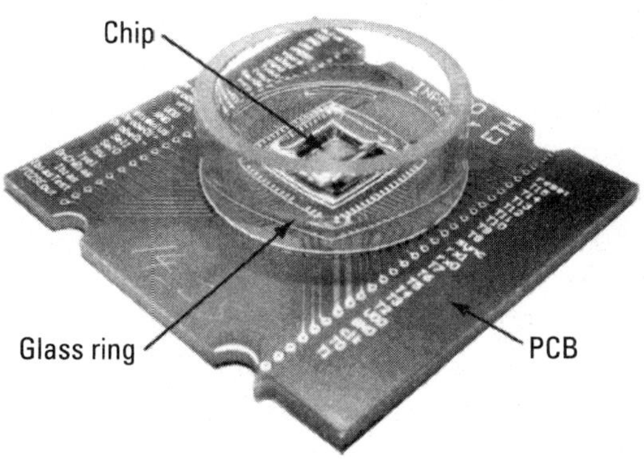

Figure 6.23 A packaged chip. (*From:* [20]. Reprinted with permission.)

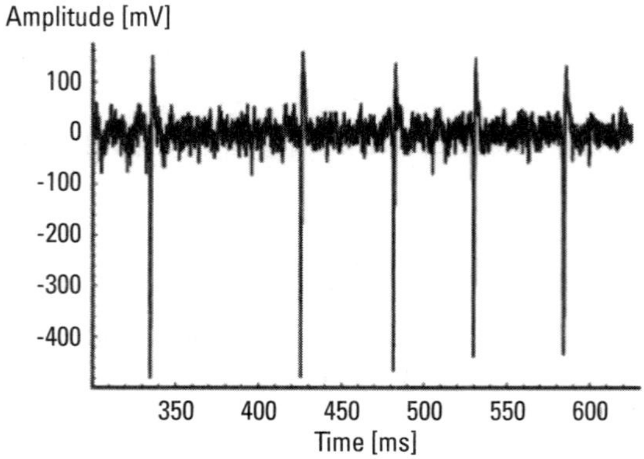

Figure 6.24 An extracellular recording from spontaneously firing chicken neurons. (*From:* [20]. Reprinted with permission.)

6.6 CMOS Lab-on-a-Chip for Cytometry Applications

As a result of research efforts in the late 1960s and 1970s, conventional flow cytometry is considered a mature technology. In flow cytometry, single particles (biological cells) in an aqueous media that have been suitably tagged with fluorescence markers are hydrodynamically focused and made to traverse a small region of space across which a focused laser beam is shone [21]. Depending on the optical characteristics of the cell traversing the beam, and the chemical markers it carries, the incident laser light will be scattered and/or fluorescence generated light will be emitted. The optical signals emitted from the particles are then collected as spectral bands of predominantly visible light. Using chromatic filters, photomultipliers, and analog-to-digital (A/D) conversion, the acquired signals can be used to identify and quantify the biophysical or biochemical characteristics of the cell sample population. The advantage of flow cytometry is that a very large number of particles can be evaluated in a very short time period. For example, systems can analyze particles at rates up to 100,000 particles per second. Moreover, this technology can be employed to physically separate single particles/cells from a heterogeneous population, which results from the principle of cell sorting in flow cytometry. Various topologies utilizing passive microchannels, waveguides, fiber optics, and/or integrated photodetectors have been demonstrated to measure forward- or side-scattered radiation as particles moving through microfluidic channels interact with light beams [22–26]. These topologies seek an alternative to conventional microscopy

Two near-field optical sensors in a 1-m bipolar process for particle shape-based flow cytometry measurements were reported by Nieuwenhuis [27]. The first device consists of a two-photodiode strip sensor that requires particles to pass over the geometric center of the structure. The second device consists of a 2 × 20 photodiode array, which, although capable of greater resolution, is plagued by a high I/O pad count and is further challenged by the vast amount of unconditioned analog data that must be processed off chip in real time. Thrush has described early work on an integrated VCSEL emitter, photodiode, and lens system suitable for coupling to standard microfluidic chips [28]. Lee et al. have reported a platform for integrated microfluidic cytometry, where optical image sensing is integrated directly on the microfluidic substrate, is employing flip-chip packaging technology [29]. This technology can be used for near-field coupling of the optical sensors to the microfluidic channels for monitoring particles (such as beads or cells), which are transported in the microchannels. A schematic diagram of the overall flip-chip digital cytometry on a chip system is illustrated in Figure 6.25 [29].

The CMOS sensor intended for optical near-field particle detection is intimately coupled to the channel in the microfluidic substrate via flip-chip on-glass assembly. The 0.18-μm CMOS sensor, designed and fabricated for near-field microfluidic integration, is shown in Figure 6.26.

The design mandates direct coupling to microfluidic channels fabricated in glass or polymer. A block diagram of the mixed-signal CMOS sensor architecture

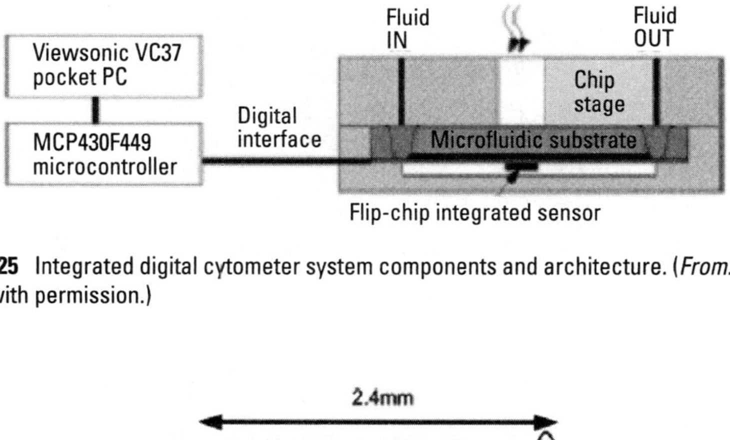

Figure 6.25 Integrated digital cytometer system components and architecture. (*From:* [29]. Reprinted with permission.)

Figure 6.26 Photograph of the linear active pixel CMOS sensor chip with interface pads, the digital logic block, and the analog signal path indicated. The sensor active area comprises a 16-element linear array. (*From:* [29]. Reprinted with permission.)

is depicted in Figure 6.27, which shows flip-chip coupling of the sensor and microfluidic substrate. It comprises a linear active pixel sensor (APS) array, correlated double sampling (CDS), and an adaptive spatial filter (SF) with a digital control block for monitoring and configuration. The sensor chip is interfaced to and controlled by an ultra-low power and battery-operable MSP430F449 microcontroller. The proposed sensor architecture, with near-field imaging capability, provides a practical basis for realizing autonomous microfluidic lab-on-a-chip cell counting platforms that are suitable for modular, mobile, and disposable applications [29].

6.7 Flip-Chip Integration

In order to validate the performance of the sensor, a family of 15 mm ×15 mm glass substrates was fabricated [29]. Each substrate consisted of two Schott D263 glass sheets thermally bonded together. A 100-μm thick sheet of glass was pat-

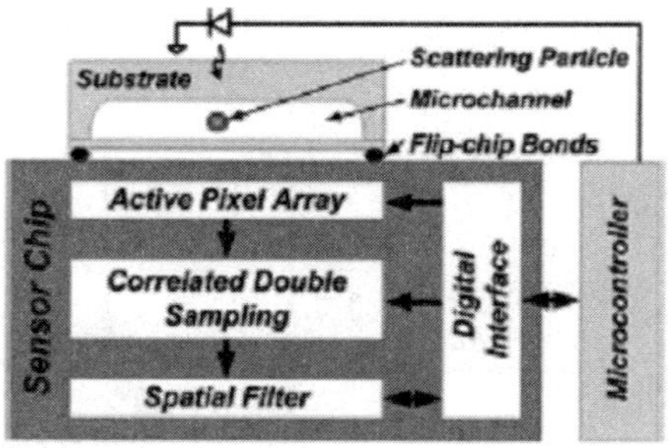

Figure 6.27 A block diagram of the mixed-signal CMOS sensor architecture. (*From:* [29]. Reprinted with permission.)

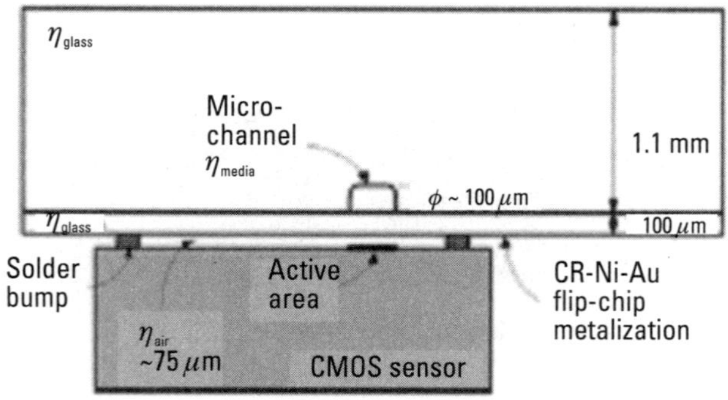

Figure 6.28 Digital cytometer hybrid substrate cross-section depicts the sensor attachment to the underside of a 100-μm-thick glass sheet that is bonded to a 1.1-mm glass sheet. (*From:* [29]. Reprinted with permission.)

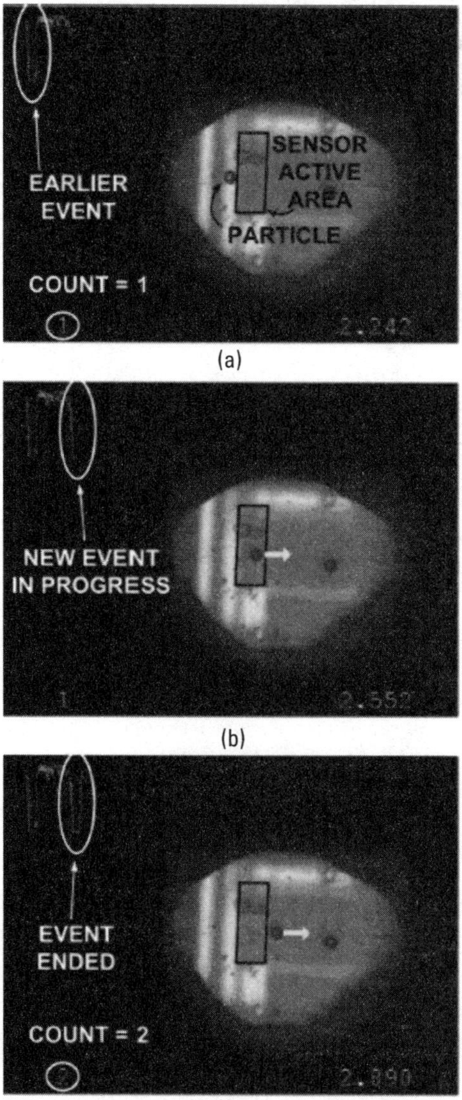

(c)

Figure 6.29 DAC output of the sensor is shown to produce a negative pulse as a 20-μm polystyrene microsphere passes across the sensor (from left to right). The real-time count of such perturbations is observed to change from 1 to 2 in frame (c). (a) Particle approaching field of view. (b) Particle in field of view (event in progress). (c) Particle exited (event ended). (*From:* [29]. Reprinted with permission.)

terned with Cr-Ni-Au (50 nm-50 nm-1,250 nm) metallization suitable for thermocompression and solder reflow flip-chip bonding. In this metal stack, nickel was employed as a barrier against chrome-gold intermetallic formation, which is known to hinder bond reliability in microelectronics packaging [30]. A 1.1-mm glass sheet delivering mechanical stability, microfluidic channels, and top-side fluid ports was then permanently bonded to the thin glass bottom sheet. This mechanical stack (including the flip-chip attached sensor) is illustrated in Figure 6.28. The glass substrates represented in Figure 6.28 contain a 100-μm-wide/deep microchannel passing transversely over the center of the sensor's linear active pixel array.

In Figure 6.29, three sampled video frames acquired by a digital video camera monitoring the sensor surface during operation are displayed along with an analog overlay of the aforementioned q-format digital value as a 16-μm diameter polystyrene bead passes over the flip-chip mounted sensor's active area (outlined in each frame). In the sequence, the particle can move from left to right over the active area of the sensor. In the top left of each frame, the synchronized digital-to-analog converter output can generate a negative pulse in response to the particle's passage over the sensor's active area. In the bottom left corner, the computed particle count is displayed and can be updated from 1 to 2 in Figure 6.29(c) as the negative pulse rises again above a static threshold.

References

[1] Hierlemann, A., et al., "Microfabrication Techniques for Chemical/Biosensors," *Proc. IEEE*, Vol. 91, No. 6, June 2003, pp. 839–863G.

[2] Medoro, N., et al., "A Lab-on-a-Chip for Cell Detection and Manipulation," *IEEE Sensors Journal*, Vol. 3, No. 3, 2003, pp. 317–325.

[3] Manaresi, N., et al., "Microelectronics Meets Biology: Challenges and Opportunities for Functional Integration in Lab-on-a-Chip," *Proceedings of the 28th Solid-State Circuits Conference (ESSCIRC 2002)*, 2002, pp. 37–42.

[4] Romani, A., et al., "A System-on-a-Programmable-Chip for Real-Time Control of Massively Parallel Arrays of Biosensors and Actuators," *Proceedings of International Conference on MEMS, NANO and Smart Systems (ICMENS 2003)*, Calgary, Alberta, Canada, 2003, pp. 236–241.

[5] Medoro, G., et al., "CMOS-Only Sensor and Manipulator for Microorganisms," *IEDM*, 2000, pp. 415–418.

[6] Ting, L., *Viscous Vertical Flows*, New York: Springer, 1991.

[7] Romani, A., et al., "Capacitive Sensor Array for Localization of Bioparticles in CMOS Lab-on-a-Chip," *IEEE International Solid-State Circuits Conference (ISSCC 04)*, 2004, pp. 456–457.

[8] Medoro, G., et al., "A Lab-on-a-Chip for Cell Detection and Manipulation," *Proceedings of IEEE Sensors*, Vol. 1, Orlando, FL, 2002, pp. 472–475.

[9] Borgatti, M., et al., "Dielectrophoresis (DEP) Based 'Lab on-a-Chip' Devices for Efficient and Programmable Binding of Microspheres to Target Cells," *Int. J. Oncology*, Vol. 27, No. 3, September 2005, pp. 1559–1566.

[10] Medoro, G., et al., "Lab on a Chip for Live-Cell Manipulation," *IEEE Design & Test of Computers*, Vol. 24, No. 1, 2007, pp. 26–36.

[11] Manaresi, N., et al., "A CMOS Chip for Individual Cell Manipulation and Detection," *IEEE Journal of Solid State Circuits*, Vol. 38, No. 12, December 2003, pp. 2297–2305.

[12] Manaresi, N., et al., "New Perspectives and Opportunities from the Wild West of Microelectronic Biochips," *Proceedings of the Design, Automation and Test in Europe Conference and Exhibition (DATE'05)*, Vol. 2, 2005, pp. 1092–1093.

[13] Romani, A., et al., "Beyond the Microscope: Embedded Detectors for Cell Biology Applications," *IEEE International Symposium on Circuits and Systems (ISCAS 2005)*, Kobe, Japan, 2005, pp. 2911–2914.

[14] Rabarot, M., et al., "Thick SU8 Lithography for BioMEMS," *Photonics West, Micromachining and Microfabrication, SPIE Conf.*, San Jose, CA, January 2003.

[15] Nicholls, J. G., A. R. Martin, and B. G. Wallace, *From Neuron to Brain*, 3rd ed., Sunderland, MA: Sinauer Associates, 1992.

[16] Molleman, A., *Patch Clamping*, New York: John Wiley & Sons, 2002.

[17] Purves, R. D., *Microelectrode Methods for Intracellular Recording and Ionophoresis*, San Diego, CA: Academic Press, 1981.

[18] Eversmann, B., et al., "A 128 × 128 CMOS Biosensor Array for Extracellular Recording of Neural Activity," *IEEE Journal of Solid State Circuits*, Vol. 38, No. 12, December 2003, pp. 2306–2317.

[19] Eversmann, B., et al., "CMOS Sensor Array for Electrical Imaging of Neuronal Activity," *IEEE International Symposium on Circuits and Systems (ISCAS 2005)*, Kobe, Japan, 2005, pp. 3479–3482.

[20] Heer, F., et al., "CMOS Microelectrode Array for Bidirectional Interaction with Neuronal Networks," *IEEE Journal of Solid State Circuits*, Vol. 41, No. 7, July 2006.

[21] Wang, Z., "Measurements of Scattered Light on a Microchip Flow Cytometer with Integrated Polymer Based Optical Elements," *Lab. Chip*, Vol. 4, No. 4, 2004, pp. 372–377.

[22] Fu, A. Y. -C., "Microfabricated Fluorescence-Activated Cell Sorters (μFACS) for Screening Bacterial Cells," Ph.D. thesis, California Institute of Technology, Pasadena, CA, 2002.

[23] Fu, L. M., "Electrokinetically Driven Micro Flow Cytometers with Integrated Fiber Optics for On-Line Cell/Particle Detection," *Analytica Chimica Acta.*, Vol. 507, No. 1, 2004, pp. 163–169.

[24] Singh, K., "Analysis of Cellular Structure by Light Scattering Measurements in a New Cytometer Design Based on a Liquid-Core Waveguide," *Proc. IEE Nanobiotechnol.*, Vol. 151, No. 1, 2004, pp. 10–16.

[25] Namasivayam, V., "Advances in On-Chip Photodetection for Applications in Miniaturized Genetic Analysis Systems," *J. Micromech. Microeng.*, Vol. 14, No. 1, 2004, pp. 81–90.

[26] LeMinh, P., "Novel Integration of a Microchannel with a Silicon Light Emitting Diode Antifuse," *J. Micromech. Microeng.*, Vol. 13, No. 3, 2003, pp. 425–429.

[27] Nieuwenhuis, J. H., "Near-Field Optical Sensors for Particle Shape Measurements," *IEEE Sensors J.*, Vol. 3, No. 5, May 2003, pp. 646–651.

[28] Thrush, E., "Integrated Semiconductor Vertical-Cavity Surface-Emitting Lasers and PIN Photodetectors for Biomedical Fluorescence Sensing," *IEEE J. Quantum Electron.*, Vol. 40, No. 5, May 2004, pp. 491–498.

[29] Lee Hartley, K., V. I. S. Kaler, and O. Yadid-Pecht, "Hybrid Integration of an Active Pixel Sensor and Microfluidics for Cytometry on a Chip," *IEEE Trans. on Circuits and Systems—I*, Vol. 54, No. 1, January 2007, pp. 99–110.

[30] Harman, G. G., *Reliability and Yield Problems of Wire Bonding in Microelectronics*, Reston, VA: Int. Soc. Hybrid Microelectron., 1991.

7

CMOS Electric-Field-Based Lab-on-a-Chip for Cell Characterization and Detection

The CMOS labs-on-a-chip presented in Chapter 6 appear to be the first PCB and CMOS labs-on-a-chip that contain actuation as well as sensing components at the cell level, and they can trap, concentrate, and quantify biocells.

The disadvantages of these CMOS lab-on-a-chips can be summarized as follows [1]:

1. Based on these labs-on-a-chip, the position of the levitated cells can be detected. However, the actual intensity of the nonuniform electric field that produces the force cannot be sensed.
2. The measurements here are indirect. In other words, there is no real-time detection of the cell response under the effect of the nonuniform electric field, since the actuation and its sensing junctions cannot occur together.
3. The sensing junction in those labs-on-a-chip depends on the inertia of the levitated cells. In other words, this sensing approach depends on an external factor, which is the inertia of the levitated cells. Thus, only cells with higher inertia can be sensed and detected by using these two labs-on-a-chip.

To overcome the previous disadvantages, Ghallab et al. proposed a CMOS lab-on-a-chip using a standard CMOS 0.18-μm technology [2–9]. Figure 7.1 shows the layout of the proposed lab-on-a-chip, with a total die area of 0.75 mm × 0.6 mm. The proposed lab-on-a-chip consists of two main parts:

1. The actuation part, which is a quadrapole electrode configuration, generates the required nonuniform electric field pattern (i.e., DEP force) to passively levitate the cell that needs to be characterized, without feedback.
2. The sensing part, a novel CMOS electric field sensor that is referred to as differential electric field sensitive field effect transistor (DeFET), senses the electric field intensity gradient above it. Thus, a real time change in the electric field due to the presence of the levitated cell can be detected. To obtain an image of the electric field, and consequently characterize the levitated cell, the DeFET in an array form is used.

7.1 Design Flow

In order to generate a successful design, the following design procedure has been followed [10]:

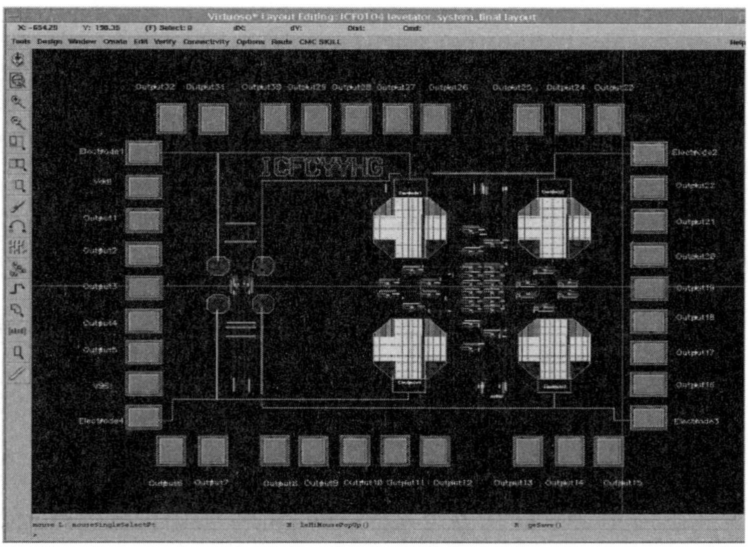

Figure 7.1 Layout of the CMOS lab-on-a-chip for cells characterization and detection.

1. *Theory:* A novel idea exists that can be implemented in CMOS technology. First, the idea needs to be verified theoretically using the basic knowledge of physics, mathematics, and circuit analysis.
2. *Physical information:* Since a decision has been made to use the CMOS 0.18-μm technology then the technology has to be checked and its important physical parameters have to be extracted, such as the doping density, gate oxide width, well depth, and drain and source depth.
3. *PISCES software:* The minimum electric field required on the gate to create a channel in a CMOS 0.18-μm transistor needs to be checked. To do this, the PISCES software has been used along with appropriate 0.18-μm CMOS parameters to check the minimum electric field required above the gate to create a channel. PISCES is software provided by Stanford University, and can be used to perform the simulation at the physical layer level [11].
4. *Design the electrodes:* From PISCES, the required electric field intensity supported by the gate to create a channel can be determined reasonably accurately. Thus, the electrodes that can provide the required electric field can be designed. Designing the electrodes includes selecting the best electrode dimensions, and determining the distance between them. At this step, the best locations for the sensors to be inserted to measure the lowest electric field need to be determined. Moreover the supply voltage required to be applied to the electrodes needs to be determined to obtain the required electric field intensity. To design the electrodes, electrostatic simulation software that can solve Maxwell equations is needed (Coulomb ver. 6.1 is used) [12].
5. *Design and model the electric field sensor (DeFET):* A CMOS CAD tool has to be used to design and model the sensor; The Cadence analog kit has been used for the sensor design. The Cadence analog kit contains three main steps: generate the schematic representation of the circuit or design; simulate the design in DC, AC, or transient mode of operations; design the layout using the layout tool and extract the parasitic, which is associated with the design.
6. *Layout of the lab-on-a-chip:* Using Cadence, the layout of the complete design of the lab-on-a-chip has been generated using Taiwan Semiconductor Microfabrication Company (TSMC) 0.18-μm technology. The lab-on-a-chip contains both the electrodes (actuation part) and the sensors (sensing part).
7. *Fabrication:* After finishing the layout, the chip was fabricated using standard TSMC 0.18-μm fabrications technology.

8. *Testing and verification:* After receiving the chip, it was tested in two steps. The first step was to test the electrical properties of the DeFET sensor. After confirming the performance of the sensor, the CMOS lab-on-a-chip, which contains an array of the sensors, was tested targeting biomedical analysis applications. In addition to testing the electrical properties of the sensor, such as its DC and AC response, the Coulomb software was used to confirm the measurements in order to be sure that the sensor gives correct readings with different electric field profiles.
9. *Design and test the readout circuit:* In parallel with designing the proposed imager, the readout circuit needs to be designed and tested, and then it needs to be tested with the lab-on-a-chip. Figure 7.2 shows a summary of the design flow steps.

7.2 Actuation

As described earlier in Chapter 3 (Section 3.1), the quadrupole electrode configuration provides a higher force value with respect to the dipole configuration [see (3.22) and (3.24)]. Therefore, the small particles can be levitated using this configuration [13–17]. Also, the quadrupole levitator comprises an azimuthally symmetric electrode arrangement capable of sustaining passive stable particle levitation [13, 14]. Moreover, as a diagnostic tool, quadrupole levitation offers researchers insight into the detailed electrical composition of materials [16, 17]. For these reasons, the quadrupole electrode configuration has been selected as an actuation part in the proposed CMOS lab-on-a-chip.

In Chapter 3, a quadrupole electrode configuration is approximated by a system of four point charges ($\pm Q$) situated in the x-y plane and arranged symmetrically about the z-axis (see Figure 3.8) [16]. To implement the same quadrupole system in the 0.18-μm CMOS technology, four identical octagons using the metal2 layer are being used. These octagons are in the x-y plane and arranged symmetrically about the z-axis (see Figure 7.1), with a distance of 100 μm between each other, as shown in Figure 7.3. Figure 7.4 shows a schematic diagram of a single electrode, the dimension of the electrode is 100 μm $\times$ 100 μm from edge to edge. This dimension actually violates the direct rule check (DRC) for the standard CMOS 0.18-μm technology, where the maximum metal area is required to be smaller than 35 μm $\times$ 35 μm. Thus, a grid arrangement of metal2 with a 1-μm space between each metal2 rectangle is used, as shown in Figure 7.4.

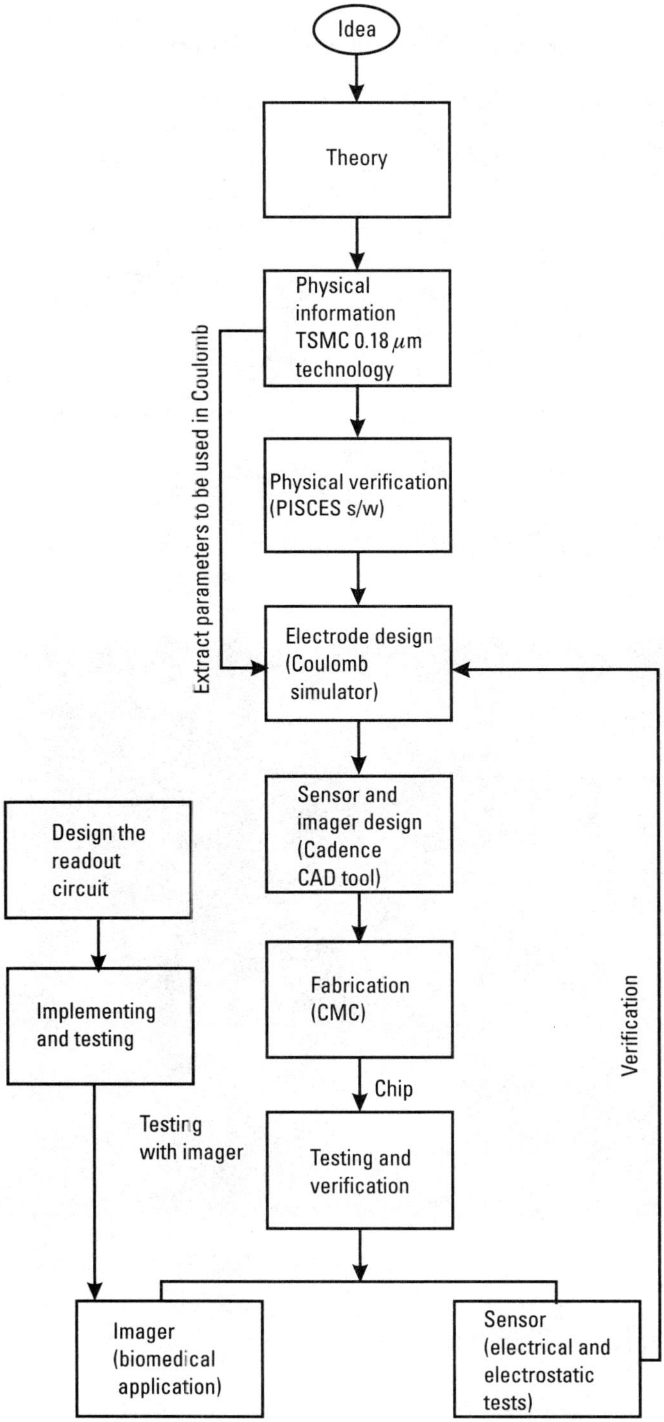

Figure 7.2 Design flow steps.

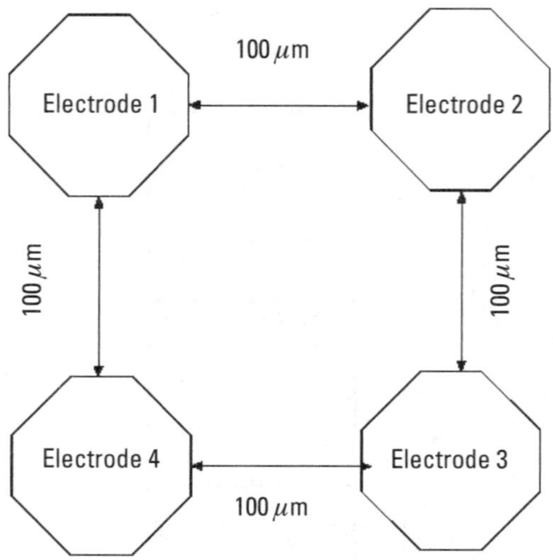

Figure 7.3 The quadrupole configuration.

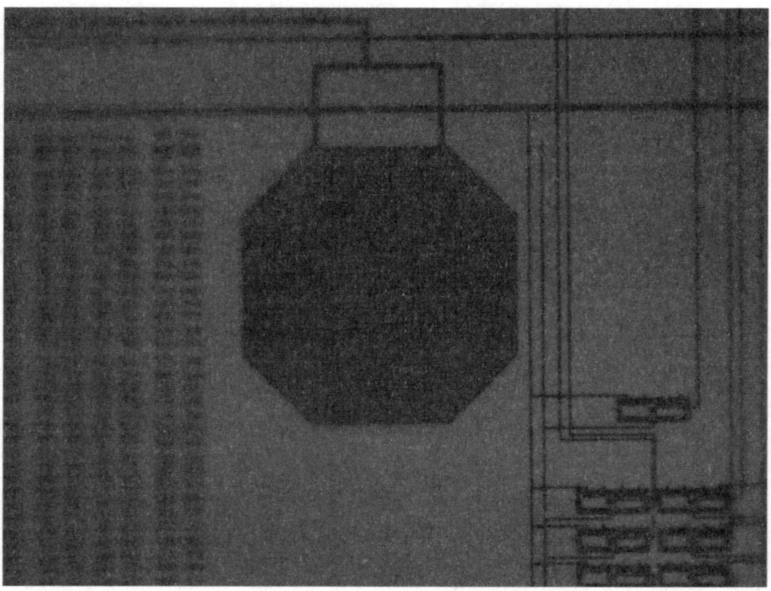

Figure 7.4 Picture of a single quadrupole electrode using metal2.

7.3 Electrostatic Simulation

Electrostatic field simulation encompasses the computer-aided engineering task of numerically solving Maxwell's equations for arbitrary geometric problems. Electrostatic simulators have gained acceptance for their ability to solve problems too complex for analytical representation. Several different numerical methods form the computational engines of various commercial solvers. For numerical simulation, a boundary element method software package was utilized. The simulation program is a 3D electrostatic design package entitled Coulomb, version 6.1; it is one of a set of boundary element method simulation packages offered by Integrated Engineering Software of Winnipeg, MB, Canada.

The boundary element method, as with any numerical method, must also discretize the problem space into elemental units. The boundary element method only discretizes the boundaries between neighboring media and then solves Maxwell's equations in integral form along those boundaries. The unknowns being solved for are physical charges and currents adequate to maintain all boundary conditions prescribed by the geometry. The equivalent sources solved for in boundary element analysis, due to their physical significance, provide intuitively simple means for direct computation of global quantities such as force, torque, stored energy and capacitance. Additionally, scalar and vector field solutions emerge from integral operations, which minimize errors by smoothing discretization noise in the boundary sources. Figure 7.5 shows the input file for the Coulomb electric field simulator. It contains the quadrupole octagon aluminum electrodes; their centers are placed at [$(100\mu, 100\mu, 0)$, $(-100\mu, 100\mu, 0)$, $(100\mu, -100\mu, 0)$, and $(-100\mu, -100\mu, 0)$], respectively, in a universe field with water, which has a relative permittivity of 80. The electrodes located at [$(100\mu, 100\mu, 0)$, $(-100\mu, -100\mu, 0)$], have an applied voltage of +3V. The other two electrodes, located at [$(-100\mu, 100\mu, 0)$, $(100\mu, -100\mu, 0)$], have an implied voltage of –3V. Six test particles of relative permittivity 9 and 20, respectively, and radii 5 μm, 10 μm, 20 μm, 30 μm, 40 μm, and 50 μm were individually positioned on the structure along the *z*-axis at heights ranging from 5–300 μm in 5-μm increments.

Coulomb was run in the DC permittivity mode to eliminate conductivity and reduce the polarization factor [K_2, see (3.26)] to a real number. At

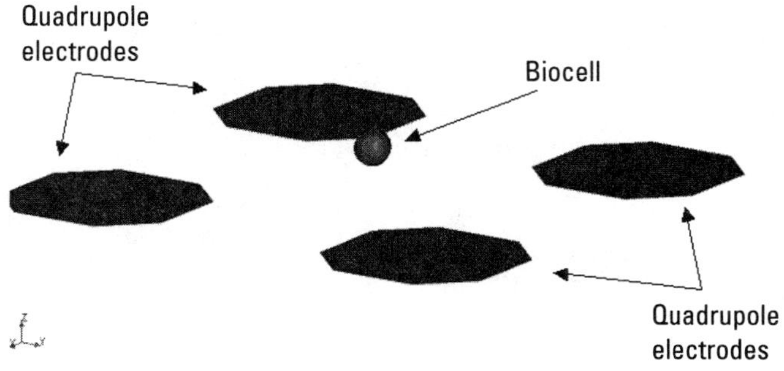

Figure 7.5 The input file for the Coulomb program.

each z-location, the net body force acting on the test particle was computed. Figures 7.6 and 7.7 show the force acting on the test particles. From these figures we notice that the force increases with the increasing radius of the particle, as seen in (3.24). Also we notice that by using the octagon electrode configuration, we can get similar force profiles to the point charge quadrupole configuration [15].

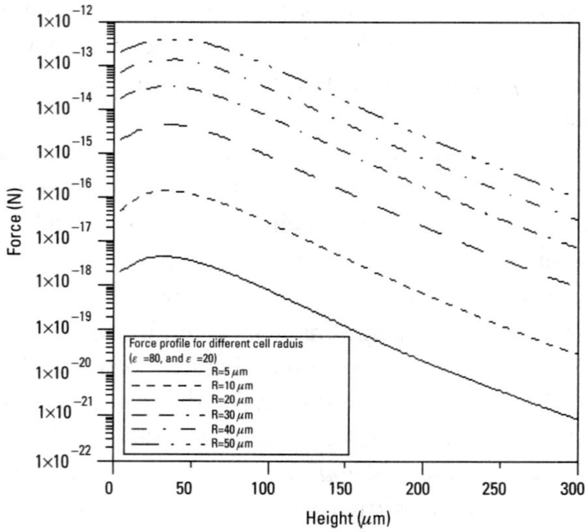

Figure 7.6 The force profile with different radii: $\varepsilon_m = 80$ and $\varepsilon_c = 9$.

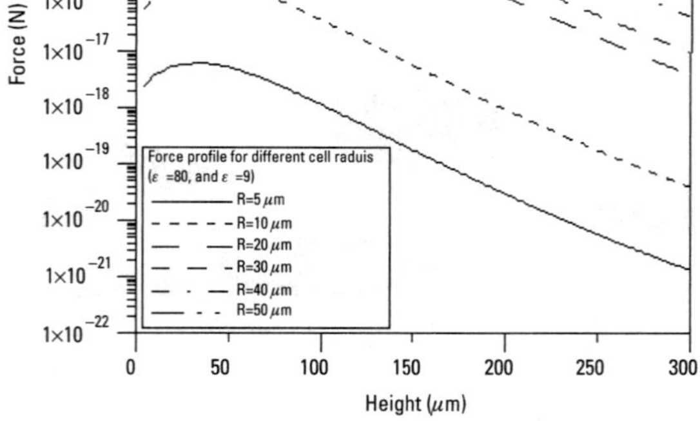

Figure 7.7 The force profile with different radii: $\varepsilon_m = 80$ and $\varepsilon_c = 20$.

7.4 Sensing

In the DEP process, the manipulating electric field is a nonuniform electric field, (i.e., the electric field is a function of the distance). Thus, we can detect the electric field by using the electric field sensitive field effect transistor (eFET), as a novel electric field sensor. To extend the range of measurement of the eFET, the CMOS concept to implement the differential electric field sensitive field effect transistor (DeFET) sensor is used [2]. The DeFET consists of two complementary eFETs, one of them is a p eFET type and the second is an n eFET type.

In the proposed lab-on-a-chip, the sensing part is composed of an array of the DeFET). This array is located around the central point of the quadrupole electrodes, where the cell will be levitated, so that we can get accurate information not only about the characteristics of the cell, but also about the intensity of the applied nonuniform electric field.

7.5 The Electric Field Sensitive Field Effect Transistor (eFET)

Figure 7.8 shows the physical structure of the eFET. It consists of two adjacent drains, two adjacent floating gates with distance *d* between them, and a single source. For the eFET, it is equivalent to two identical enhancement MOSFET devices, as shown in Figure 7.9. Thus, the two drain currents are equal if a uniform electric field is applied.

Under the influence of the nonuniform electric field, the current imbalances of the two drain currents occur. Due to the drain current dependence on the gate voltage, the eFET device that has two adjacent gates, and two adjacent drains, but isolated from each other, can sense the difference between the two gate voltages, which reflects the gradient of the applied nonuniform electric field. Figure 7.10 shows the eFET's circuit symbol. To increase the dynamic range of

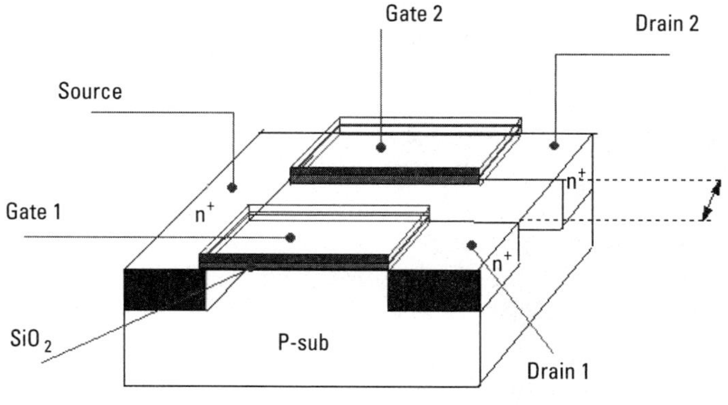

Figure 7.8 Physical structure of an eFET. (*From:* [2]. Reprinted with permission.)

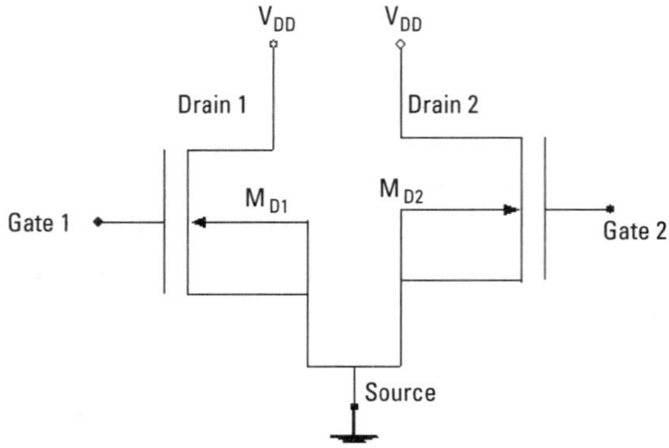

Figure 7.9 Equivalent circuit of an eFET. (*From:* [2]. Reprinted with permission.)

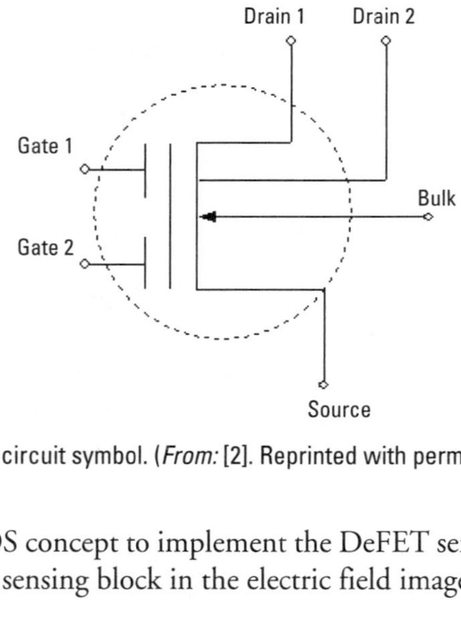

Figure 7.10 The eFET's circuit symbol. (*From:* [2]. Reprinted with permission.)

the eFET, the CMOS concept to implement the DeFET sensor is used; this sensor will be the basic sensing block in the electric field imager.

7.6 The Differential Electric Field Sensitive Field Effect Transistor (DeFET)

The DeFET consists of two complementary eFETs, one of them is a p eFET type and the second is an n eFET type, as shown in Figure 7.11. The equivalent circuit of the DeFET is shown in Figure 7.12. From Figure 7.12, the two gates of the p eFET and the n eFET are connected with each other, and there is a cross coupling at the output between the two drains of the p eFET and the n eFET. The output current I_O is equal to the difference between the two drain currents

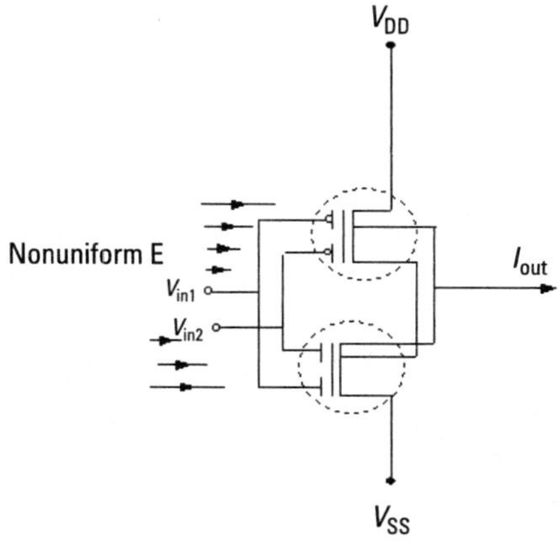

Figure 7.11 The DeFET's circuit symbol. (*From:* [2]. Reprinted with permission.)

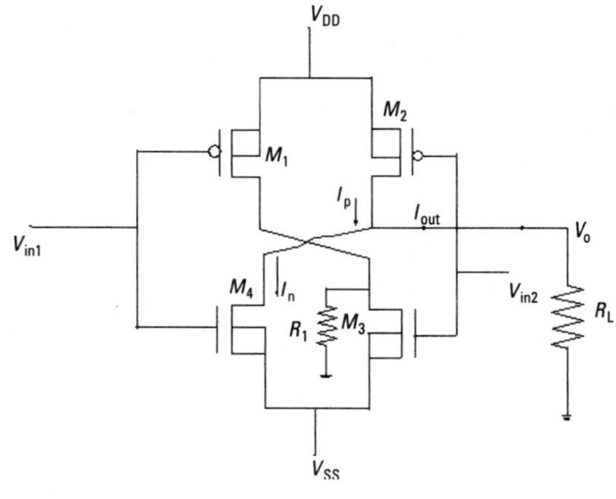

Figure 7.12 An equivalent circuit of a DeFET. (Reprinted from [2] with permission.)

$I_p - I_n$ (i.e., $I_O = I_p - I_n$, see Figure 7.12); where, I_p and I_n are functions of the two applied gate voltages V_{in1} and V_{in2}, respectively. In DeFET's design, by adjusting the dimensions (the width and length) of M_2 and M_4, I_O is directly related to the difference between the two applied gate voltages ($V_{in1} - V_{in2}$), and since $V_{in1} - V_{in2}$ is equal to the applied electric field above the two gates multiplied by the distance between them $\left(\frac{V_{in1} - V_{in2}}{d} = E\right)$, where d is the distance between the two split gates, which is constant. So, I_O is related directly to the intensity of the applied nonuniform electric field. Thus by measuring I_O, and consequently V_o, the intensity of the nonuniform electric field, can be detected.

7.7 DeFET Theory of Operation

From Figure 7.12, a simple analysis shows that the output current (I_{out}) is:

$$I_{out} = I_p - I_n \tag{7.1}$$

where I_p and I_n are the two drain currents in the p transistor (M_2) and the n transistor (M_3), respectively. The sensitivity (S_1) is given by

$$S_1 = \frac{dI_{out}}{dE} \tag{7.2}$$

Also, the gradient in the electric field intensity above the sensor (E), which is modified by the presence of the sensor and produces ΔV, is related to this voltage as follows:

$$\Delta V = -E.d \tag{7.3}$$

where ΔV is the difference between the two gate input voltages (i.e., $\Delta V = V_{in1} - V_{in2}$).

From (7.2), I_{out} can be expressed as:

$$I_{out} = \int S_1 d(E) \tag{7.4}$$

$$I_{out} = S_1 E + \text{Constant} \tag{7.5}$$

and V_{out} is:

$$V_{out} = I_{out} R_L = S_1 R_L E + \text{Constant} \tag{7.6}$$

where R_L is the load resistance. Thus, we can rewrite (7.6) as:

$$V_{out} = SE + \text{Constant} \tag{7.7}$$

where

$$S = S_1 R_L \tag{7.8}$$

To get the value of the "Constant" in (7.7), we use the condition $E = 0$, so Constant = V_{out} when $E = 0$.

We notice that in our case, $E = 0$ means that $\Delta V = 0$, so we have the same voltage drops at the two input gates (i.e., $\Delta V = V_{in1} - V_{in2} = 0$), so $V_{in1} = V_{in2}$. $V_{in1} = V_{in2}$ means that we apply a uniform electric field. So, we can rewrite (7.7) as:

$$V_{out} = V_{Non} + V_{Uni} \tag{7.9}$$

where $V_{Non} = SE$ is the output voltage when we apply a nonuniform electric field, and V_{Uni} is the output voltage when we apply a uniform electric field.

Substituting (7.1) into (7.2),

$$S_1 = \frac{d\left(I_p - I_n\right)}{dE} \tag{7.10}$$

$$S_1 = \frac{d\left(I_p - I_n\right)}{d(\Delta V)} \cdot \frac{d(\Delta V)}{dE} \tag{7.11}$$

For simplicity, we will assume that the transconductances (g_m) of the M_2 and M_4 transistors are equal; this can be obtained by adjusting the dimensions of M_2 and M_4 to obtain

$$\left(I_p - I_n\right) = -g_m \Delta V \tag{7.12}$$

From (7.12)

$$\frac{d\left(I_p - I_n\right)}{d(\Delta V)} = -g_m \tag{7.13}$$

From (7.3)

$$\frac{d(\Delta V)}{dE} = -d \tag{7.14}$$

From (7.13) and (7.14) into (7.11), the sensitivity can be given as:

$$S_1 = g_m d \tag{7.15}$$

From (7.15) into (7.8),

$$S = g_m dR_L \tag{7.16}$$

As a linear equation, we can express the output voltage (V_{out}) in terms of the sensitivity and the electric field as follows:

$$V_{out} = SE + V_{Uni} \tag{7.17}$$

Substituting (7.16) into (7.17),

$$V_{out} = g_m dR_L E + V_{Uni} \tag{7.18}$$

Equation (7.18) shows a linear relationship between the DeFET's output voltage and the intensity of the applied electric field. Thus, if we have an array of DeFETs, then we can obtain the electric field intensity at different locations and consequently the electric field gradient.

In other words, we can rewrite (7.18) in terms of ΔV as:

$$V_{out} = -g_m R_L \Delta V + V_{Uni}\left(V_{in1}, V_{in2}\right) \tag{7.19}$$

$$V_{out} = V_{Non} + V_{Uni}\left(V_{in1}, V_{in2}\right) \tag{7.20}$$

where $\Delta V g_m R_L$ represents the output voltage as a result of a nonuniform electric field (V_{Non}).

From (7.20), there are two different output voltages. The former, V_{Non}, is linearly related to the change in ΔV (i.e., we have a nonuniform electric field applied), and the latter, V_{Uni}, is a constant, which can be calculated when $V_{in1} = V_{in2} = V_{in}$ (i.e., $\Delta V = 0$, so we have a uniform electric field applied). However, its value changes by changing V_{in}.

7.8 Modeling the DeFET

As a new device, the DeFET is not a standard device in the simulator libraries. Thus, a macromodel to evaluate the performance of circuits composed of DeFET and MOSFET devices is needed. A DC model has been proposed and tested using SPECTRE version 5. Also, an equivalent DC circuit of the DeFET using PSPICE version 9.1 has been proposed so it can be used in the SPICE environment.

7.8.1 A Simple DC Model

Figure 7.13 shows the proposed DeFET circuit used for SPECTRE simulation. The channel length and the width of the PMOS and NMOS used are 0.4 μm/10 μm and 0.4 μm/1 μm, respectively. Figure 7.14 shows the simulation results of the output voltage V_{out} against the input voltage difference $V_{in1} - V_{in2}$ (ΔV), where V_{in1} varies from −2.5V to 2.5V and V_{in2} is 0V. From Figure 7.14, the linear relationship between V_{out} and ΔV can be observed. As the external electric field $E = -\Delta V/d$, and $d = 0.5$ μm, Figure 7.15 shows the output voltage V_{out} against the intensity of the applied electric field. We can observe that V_{out} is linearly related to E, as we expected from the theory of the DeFET [see (7.20)]. The sensitivity can be determined from Figure 7.15 as $S = \Delta V_{out}/\Delta E = 0.12$ (V/V/μm). If we vary

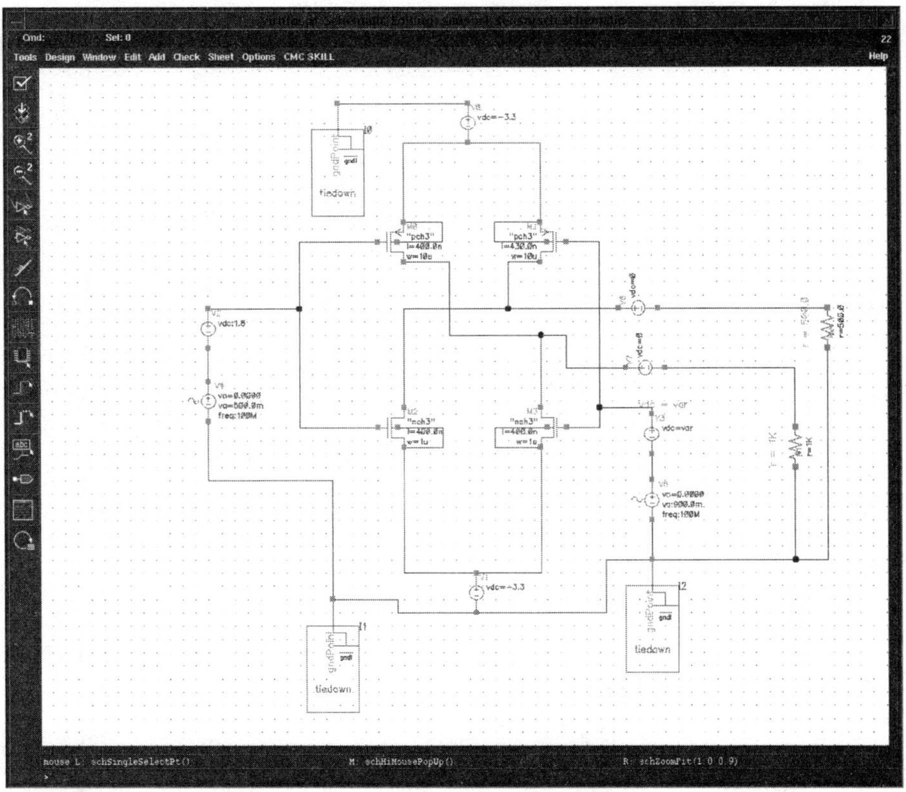

Figure 7.13 Schematic of the DeFET used in SPECTRE simulation.

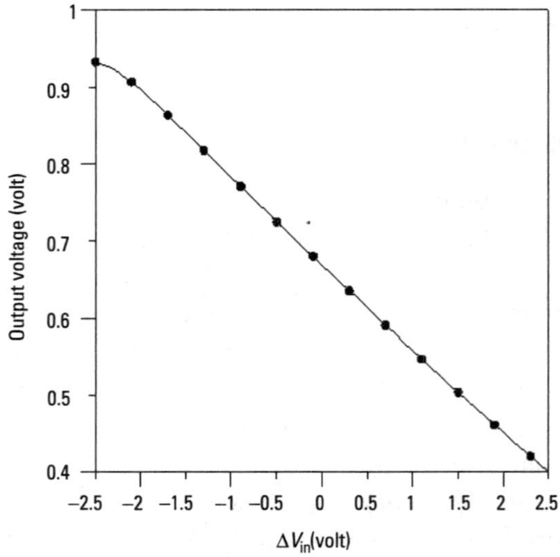

Figure 7.14 DC response of the DeFET.

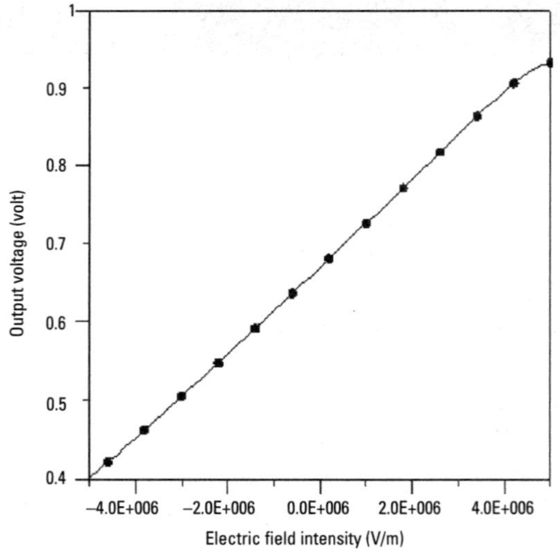

Figure 7.15 Output voltage versus electric field intensity.

V_{in1} from −2.5V to 2.5V and V_{in2} from −3V to 3V, we can get the same linear relationship between V_{out} and V_{in1} for different V_{in2} values, as shown in Figure 7.16.

7.8.2 SPICE DC Equivalent Circuit

Figure 7.17 shows the SPICE equivalent circuit of the DeFET. As outlined before, in DeFET's theory of operation (Section 7.3), there are two sources of the

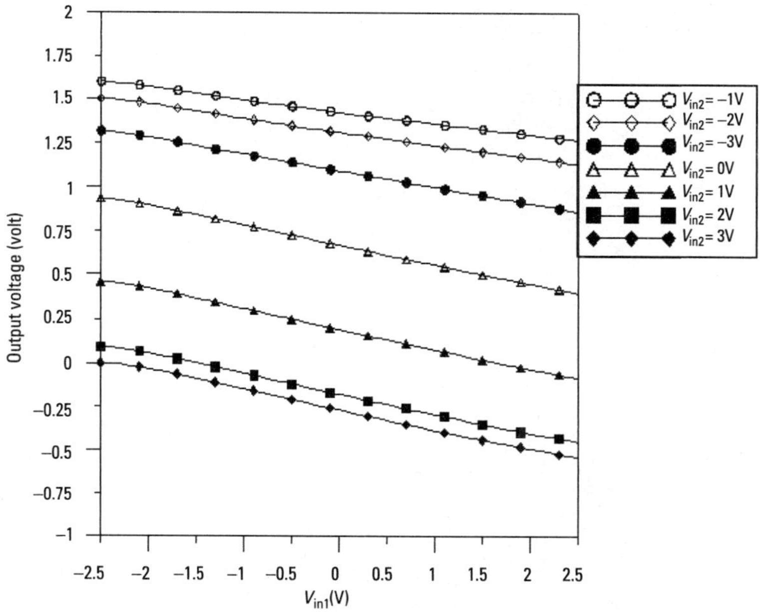

Figure 7.16 SPECTRE's simulation results.

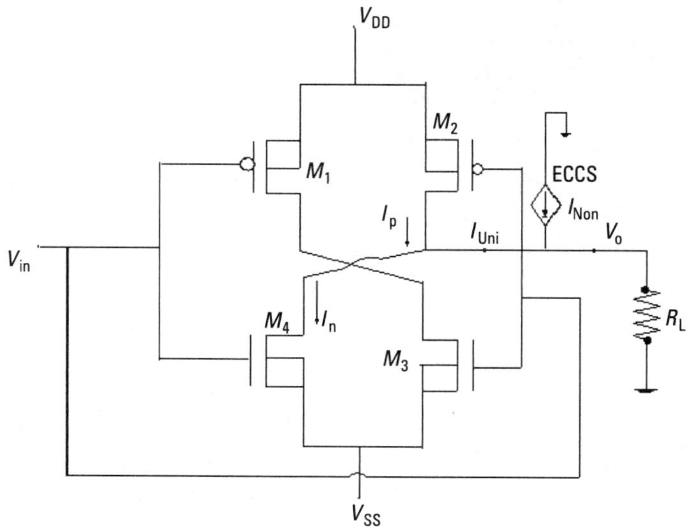

Figure 7.17 DeFET SPICE equivalent circuit. (*From:* [21]. Reprinted with permission.)

electric field [21]. The first one is a uniform electric field source, and the second is a nonuniform electric field; both of these fields can control the output voltage (V_{out}). The output current caused by the uniform electric field (I_{uni}) is represented by the four MOSFETs (M_1 through M_4) with the two gates connected

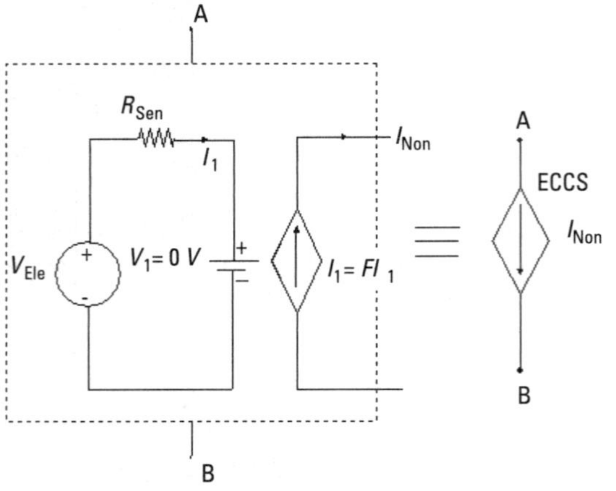

Figure 7.18 Electric field controlled current source (ECCS) model. (*From:* [21]. Reprinted with permission.)

(see Figure 7.17). I_{Uni} represents the output current when the two gates are connected. The output current caused by a nonuniform electric field (I_{Non}) is represented by an electric field controlled current source (ECCS). I_{non} is related to the electric field E, and the sensitivity S by the equation ($I_{non} = SE$). Figure 7.18 shows the SPICE equivalent circuit of ECCS. In this SPICE circuit, the external electric field applied is represented as a voltage source V_{Ele}. The value of R_{Sen} is proportional to the inverse value of the sensitivity S. I_{Non} is a current source controlled by I_1, which is caused by the nonuniform external electric field. As a simple model, Figure 7.19 shows this simple DC model, where the output current (I_{out}) is the summation of two dependent sources of current [21]. The first is an electric field controlled current source (I_{Non}), which depends on the applied nonuniform electric field, and the second is a voltage-controlled current source (I_{Uni}), which represents the output current obtained when applying a uniform electric field.

Figure 7.20 shows the schematic input file for the PSPICE, level 7 models for the PMOS and NMOS transistors is used [21]. Figure 7.21 shows the simulation results of the output voltage V_{out} against the input voltage V_3, where V_3 varies from −2.5V to 2.5V and V_4 varies from –3V to 3V. From Figure 7.21, V_3 represents the uniform electric field source, while V_4 represents the nonuniform electric field source. Also, one can observe the linear relationship between V_{out} and V_{in1} for different values of V_{in2}. In Figure 7.21, the simulation results obtained from the two different proposed models are drawn using PSPICE and SPECTRE simulators. Good agreements between these two models can be observed. Thus, both of these models can properly represent the DC response of the DeFET.

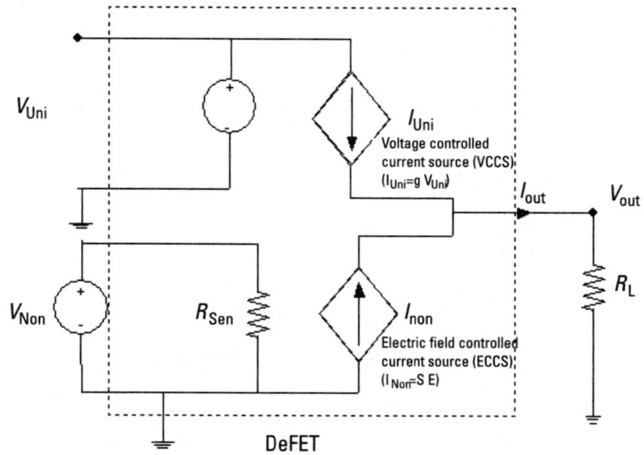

Figure 7.19 A simple DC model. (*From:* [21]. Reprinted with permission.)

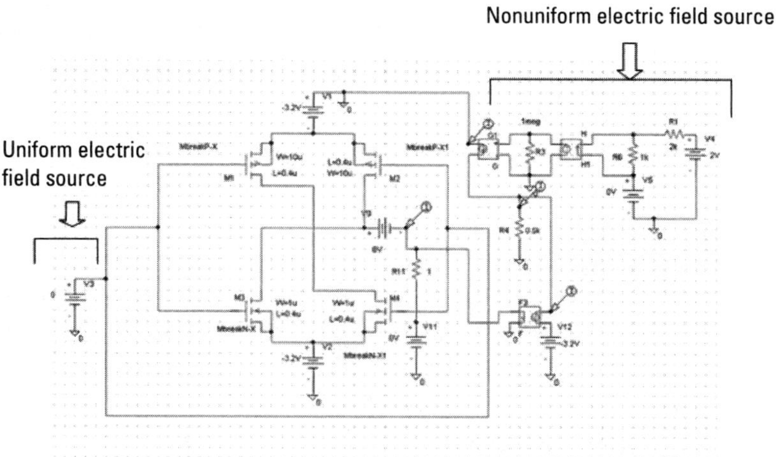

Figure 7.20 DeFET PSPICE equivalent circuit. (*From:* [21]. Reprinted with permission.)

As the external electric field $E = -\Delta V/d$, and $d = 0.5\ \mu$m, Figure 7.22 shows the output voltage V_{out} against the electric field applied. One can observe that V_{out} is linearly related to E, which is to be expected from the theory of the DeFET [see (7.20)]. The sensitivity can be determined from Figure 7.22 as $S = \Delta V_{out}/\Delta E$ = 0.12 (V/ V/μm).

7.8.3 AC Equivalent Circuit

Figure 7.23 shows the AC equivalent circuit of the DeFET, where the media is represented by a capacitor in parallel with a resistor ($C_{ext}//R_{ext}$) [18]. The media

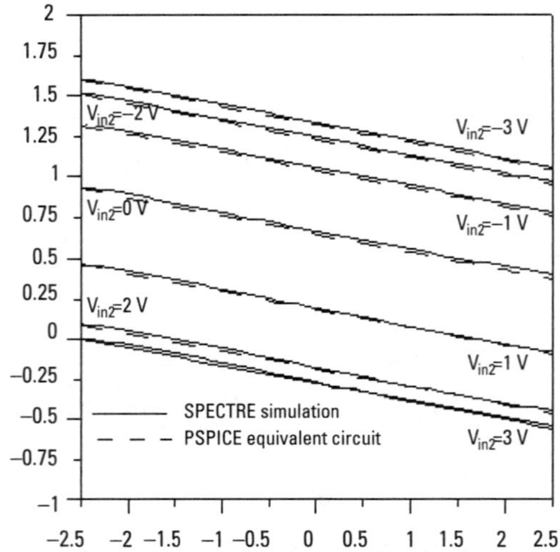

Figure 7.21 SPECTRE and PSPICE simulation. (*From:* [21]. Reprinted with permission.)

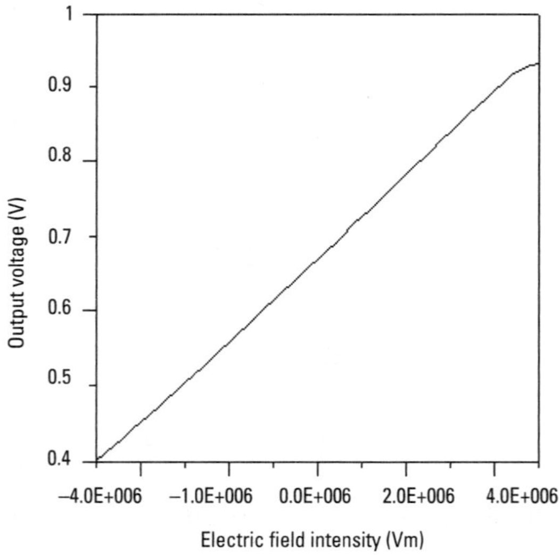

Figure 7.22 PSPICE output voltage versus electric field intensity. (*From:* [21]. Reprinted with permission.)

here can be the dielectric fluid above the DeFET sensor, that is, any neutral body, such as: biocells, DNA molecules, or a virus. The AC response for different C_{ext} values is shown in Figure 7.24. From this figure we can observe that the DeFET is working as a bandpass filter. The magnitude, bandwidth, and the frequency range of operation of this filter depend on the external media (see Figure 7.24). These factors are very important to extract useful information about the media itself. A summary of these results is shown in Table 7.1.

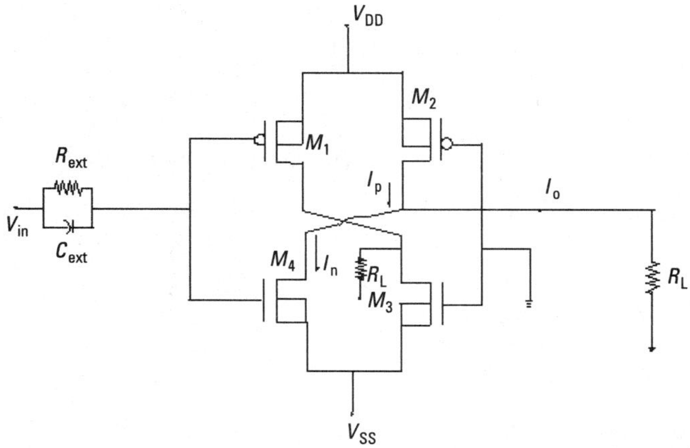

Figure 7.23 The AC equivalent circuit of the DEFET. (*From:* [2]. Reprinted with permission.)

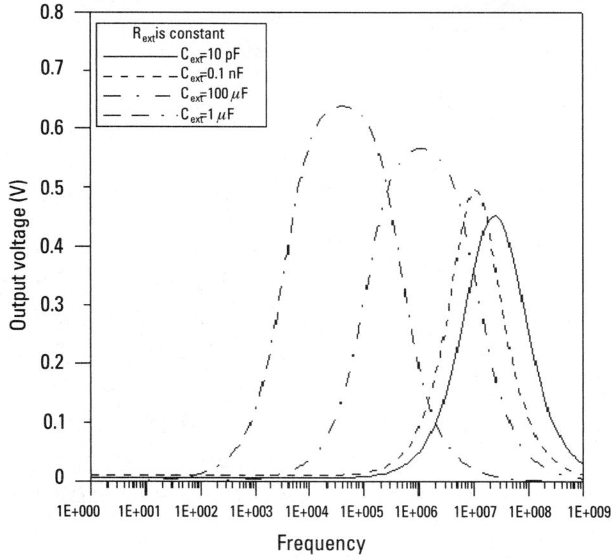

Figure 7.24 AC response with R_{ext} = 500 KΩ and variable C_{ext}. (*From:* [2]. Reprinted with permission.)

7.9 The Effect of the DeFET on the Applied Electric Field Profile

To check the effect of the DeFET on the applied electric field profile, an electrostatic simulation has been done using Coulomb software, with the same conditions that we have used before without the DeFET (refer to Section 7.3). Figure 7.25 shows the input file for the Coulomb electric field simulator. It contains the quadrupole octagon aluminum electrodes; their centers are placed at [(100μ, 100μ, 0), (−100μ, 100μ, 0), (100μ, −100μ, 0), and (−100μ, −100μ, 0)], re-

Table 7.1
Summary of the AC Response When R_{ext} = 500 KΩ and Variable C_{ext}

C_{ext}	Central Frequency (fo)	Bandwidth
10 pF	26.7 MHz	100 MHz
0.1 nF	10 MHz	26.59 MHz
100 nF	2.6 MHz	9.841 MHz
1 μf	50.1 kHz	311 kHz

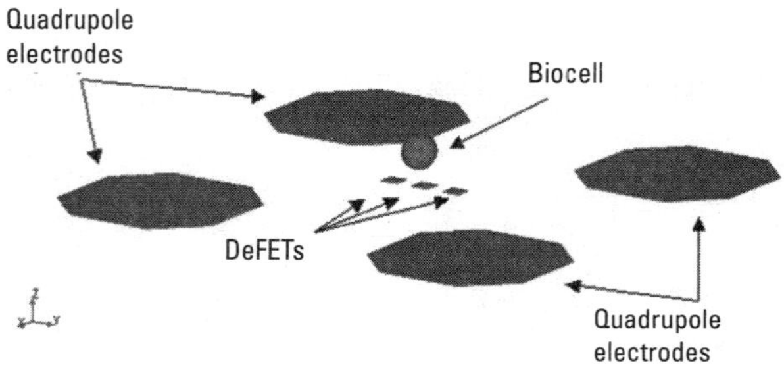

Figure 7.25 Electrostatic representation for the quadrupole, biocell, and DeFET sensors.

spectively, in water as the suspended medium (relative permittivity = 80) with the DeFET sensors inserted in between. The material, thickness, and permittivity of each layer of the DeFET are shown in Table 7.2. This information is extracted from the TSMC 0.18-μm technology file. The electrodes located at [(100μ, 100μ, 0), (−100μ, −100μ, 0)] (i.e., electrode 1 and electrode 3) had imposed an input voltage of +3V, the two electrodes located at [(−100μ, 100μ, 0), (100μ, −100μ, 0)] (i.e., electrode 2 and electrode 4) had imposed an input voltage of −3V.

Six test particles of relative permittivity 9 and 20 and radii 5 μm, 10 μm, 20 μm, 30 μm, 40 μm, and 50 μm were individually positioned into the structure along the *z*-axis at heights ranging from 5–300 μm in 5-μm increments. Coulomb was run in DC permittivity mode to eliminate conductivity and reduce the polarization factor [K_2; see (3.26)] to a real number. At each *z*-location, we compute the net body force acting on the test particle. The results with and without sensors are shown in Figures 7.26 and 7.27.

To clearly see the effect of the DeFET on the applied electric field, one of the dielectrophoretic force results has been redrawn with and without the DeFET in the same graph, as shown in Figure 7.28. Also, Figures 7.29 and 7.30 show the electrostatic contour for the electric field profile extracted from the

Table 7.2
DeFET Layers Parameters Used in Coulomb Input File

Material	Thickness	Relative Permittivity
Polysilicon	0.2 μm	1.4
Silicon oxide	50Å	13.8
Silicon substrate	2 μm	11.7

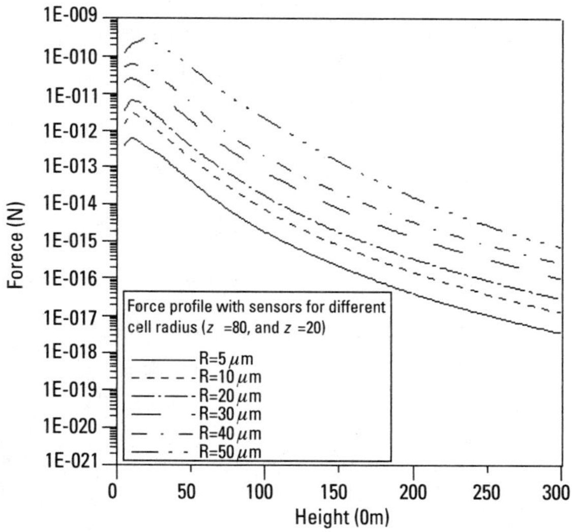

Figure 7.26 The force profile when the sensors exist with $\varepsilon_m = 80$, and the cell's relative permittivity $\varepsilon_c = 20$.

simulator with and without the DeFET sensors. From these figures the following can be observed:

- The DeFET does not disturb the profile of the applied electric field. A minimum electric field (i.e., minimum energy area) still exists in between the DeFET sensors (see Figure 7.28) and cells above the DeFETs sensors can still be levitated or trapped. We can simultaneously actuate the electric field and sense it in real time; this is a significant advantage over the proposed lab-on-a-chip in [19, 20].
- In Figure 7.28, there are two levitation areas, one of them is unstable; the slope of $\frac{dF_z}{dz}$ is positive, and the second is a stable, that is, when $\frac{dF_z}{dz}$ is negative [15]. A biocell cannot passively be levitated in the unstable re-

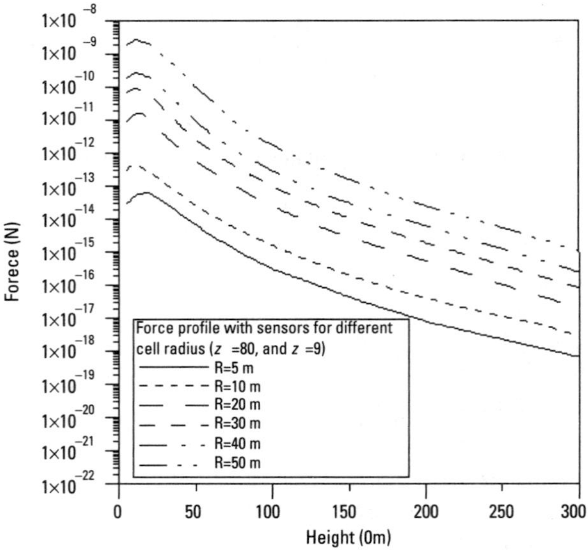

Figure 7.27 The force profile when the sensors exist with $\varepsilon_m = 80$, and $\varepsilon_c = 9$.

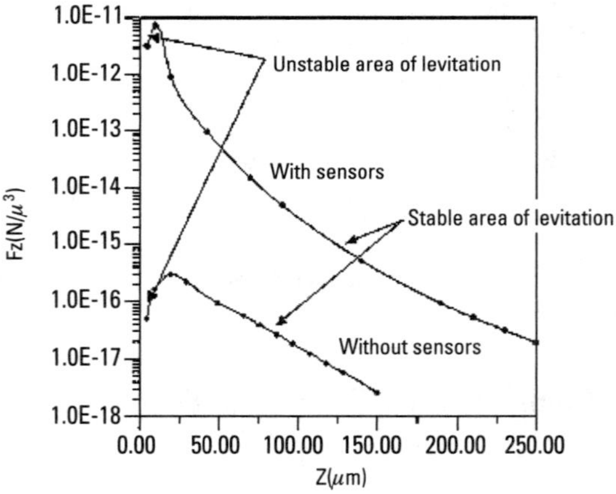

Figure 7.28 Result of the electrostatic simulation shows the improvement due to using DeFET.

gion, as an external force (feedback controlled levitation) is needed. From Figure 7.28, DeFETs improve the profile as the stable area of levitation can be achieved within a very small levitation height ($Z > 5\ \mu$m), not when $Z > 50\ \mu$m with no sensors. The insertion of the DeFETs reduces the appearance of the unstable region of operation; thus, the cells can easily be levitated passively without any external forces.

- The *z*-component of the dielectrophoretic force (see Figure 7.28) is increased so that the heavy cells can be levitated without any need for other

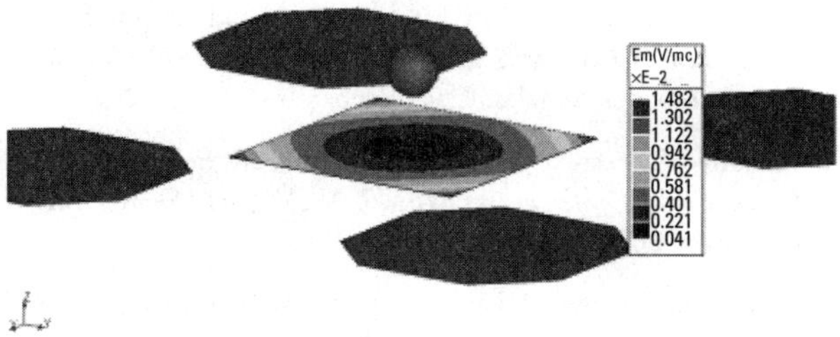

Figure 7.29 Electrostatic representation for the electrodes without sensors and the electric field intensity result.

Figure 7.30 Electrostatic representation for the electrodes with sensors and the electric field intensity result.

external forces. The cell can be levitated far from the electrodes so that many processes can be performed (e.g., cell fusion).

References

[1] Ghallab, Y. H., and W. Badawy, "Sensing Methods of Dielectrophorieses from Bulky Instruments to Lab-on-a-Chip," *IEEE Circuit and Systems Magazine*, Vol. 4, No. 3, June 2004, pp. 5–15.

[2] Ghallab, Y. H., and W. Badawy, "DeFET, a Novel Electric Field Sensor for Lab-on-a-Chip and Biomedical Applications," *IEEE Journal of Sensors*, Vol. 6, No. 4, August 2006, pp. 1027–1037.

[3] Ghallab, Y. H., and W. Badawy, "Field Imager," U.S. Patent 767235, June 2007.

[4] Ghallab, Y. H., and W. Badawy, "Field Imager," Canadian Patent Pending 2,511,898, July 2005.

[5] Ghallab, Y. H., and W. Badawy "A Single CMOS Chip for Biocell Tracking, Levitation, Detection, and Characterization," *International Symposium on Circuit and System (ISCAS 06)*, Greece, May 2006 pp. 3349–3352.

[6] Ghallab, Y. H., and W. Badawy, "On-Chip Electrical Field Sensing for Lab-on-a-Chip Applications," *Symposium of Dielectrics and the Dielectric-Electrolyte Interface in Biological and Biomedical Applications*, Los Angeles, CA, October 16–21, 2005.

[7] Ghallab, Y. H., and W. Badawy "Techniques for Biocells Sensing, Detection and Characterization," *International Conference on MEMS, NANO, and Smart Systems 2005 (ICMENS 2005)*, Banff, Canada, 2005, pp. 402–405.

[8] Ghallab, Y. H., and W. Badawy "A Lab-on-a-Chip for Biocells and DNA Analyses," *Nanotech Insight Conference 2005*, Luxor, Egypt, February 20–25, 2005, pp. 58–59.

[9] Ghallab, Y. H., and W. Badawy, "A CMOS Lab-on-a-Chip for Biomedical Applications," *International Symposium on Circuit and System (ISCAS 05)*, Kobe, Japan, 2005, pp. 1346–1349.

[10] Ghallab, Y. H., "A Novel CMOS Electric Field Imager for Lab-on-a-Chip and Biomedical Applications," Ph.D. thesis, Electrical and Computer Department, University of Calgary, Calgary, Alberta, 2005.

[11] *Manual of PISCES-2 SET and Its Applications*, Palo Alto, CA: Stanford University Press.

[12] *Coulomb Users and Technical Manual, Version 6.1*, Winnipeg, MB, Canada: Integrated Engineering Software, 2004.

[13] Hartley, L. E., K. V. I. S. Kaler, and R. Paul, "Quadrupole Levitation of Microscopic Dielectric Particles," *Journal of Electrostatics*, Vol. 46, 1999, pp. 233–246.

[14] Voldman, J., "A Microfabricated Dielectrophoretic Trapping Array for Cell-Based Biological Assays," Ph.D. thesis, Massachusetts Institute of Technology, June 2001.

[15] Jones, T. B., *Electromechanics of Particles*, Cambridge, U.K.: Cambridge University Press, 1995.

[16] Washizu, M., and O. Kurosawa, "Electrostatic Manipulation of DNA in Microfabricated Structures," *IEEE Trans. on Industry Applications*, Vol. 26, No. 6, 1990, pp. 1165–1172.

[17] Casanella, R., et al., "Aggregation Profile Characterization in Dielectrophoretic Structures Using Bacteria and Submicron Latex Particles," *IEE Proceedings on Nanobiotechnnology*, Vol. 150, 2003, pp. 70–74.

[18] Romani, A., et al., "Capacitive Sensor Array for Localization of Bioparticles in CMOS Lab-on-a-Chip," *IEEE International Solid-State Circuits Conference (ISSCC 04)*, 2004, pp. 456–457.

[19] Medoro, G., et al., "CMOS-Only Sensors and Manipulation for Microorganisms," *Proceedings of IEDM*, 2000, pp. 415–418.

[20] Medoro, G., et al., "A Lab-on-a-Chip for Cell Detection and Manipulation," *IEEE Sensors Journal*, Vol. 3, No. 3, 2003, pp. 317–325.

[21] Ghallab, Y. H., and W. Badawy, "Differential Electric Field Sensitive Field Effect Transistor: Characteristics, Modeling and Applications," *IEEE Biodevices, Funchal*, Madeira, Portugal, January 28–31, 2008.

8

Prototyping and Experimental Analysis

The lab-on-a-chip is implemented in TSMC CMOS 0.18-μm technology. The die is composed of two parts, as shown in Figure 8.1, with total die area 0.75 mm × 0.6 mm. Part I is used for testing and calibrating the differential electric field sensitive field effect transistor (DeFET) sensor, while part II is the proposed imager, which can be used to levitate and characterize biocells. Both part I and part II consist of two main pieces:

1. The actuation part, which is a quadrupole electrode configuration that produces the required nonuniform electric field profile.
2. The sensing part, which is an array of the DeFETs. The dimension of the electrodes in part I is smaller than those in part II (35 μm from edge-to-edge for part I, and 100 μm from edge-to-edge for part II).

8.1 Testing the DeFET

The fabricated DeFET is implemented in a standard CMOS TSMC 0.18-μm technology. Figure 8.2 shows a microscopic picture of two DeFETs and the electrodes used to apply the required electric field pattern. The length of each electrode is 35 μm from edge-to-edge. The dimension of the DeFET is 25 μm × 20 μm. The arrangement of the various laboratory components used during experimentation is shown in Figure 8.3.

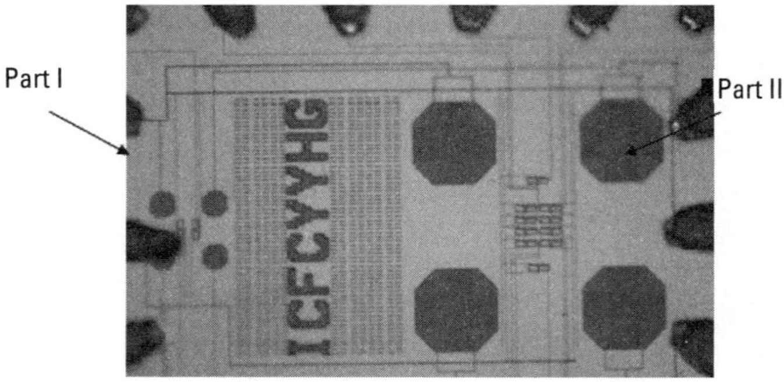

Figure 8.1 The die picture shows the two main pieces. (*From:* [12]. Reprinted with permission.)

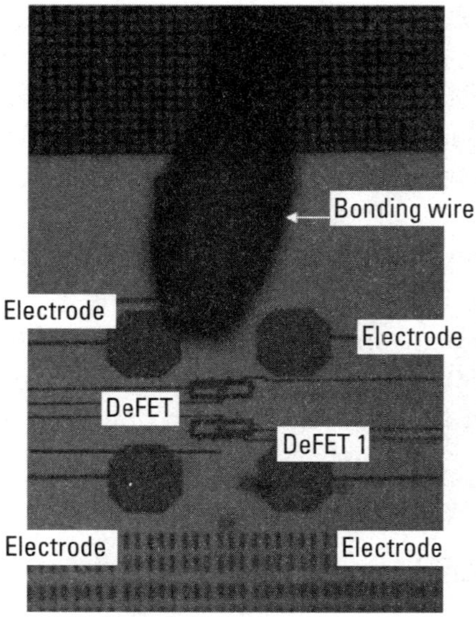

Figure 8.2 A microscopic picture shows two DeFET sensors and the electrodes used to apply the electric field. (*From:* [12]. Reprinted with permission.)

8.1.1 The DC Response

To test the DC response of the DeFET, DC voltages with different values, signs, and configurations have been applied to the four electrodes surrounding the DeFET sensors, hence varying the magnitude and sign of the applied electric field (E). DeFET 1 only has external connections to its two gates. Thus, by applying the electrode voltages and measuring the voltage drops across the two gates of DeFET 1, the gradient in the electric field intensity above DeFET 1 can be measured. At the output, the output voltage associated with each value of the measured electric field above the gates will be measured and compared with the

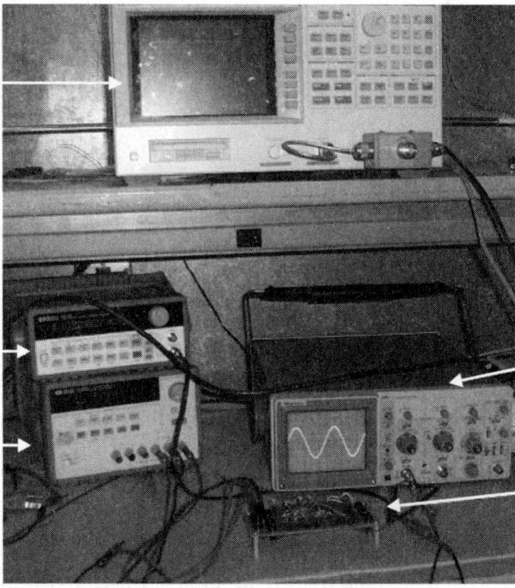

Figure 8.3 The equipment used for testing the DeFET.

simulation results, that is, using the SPECTRE circuit simulator [12]. The result is shown in Figure 8.4, from which a good agreement between the experimental and the simulation results can be observed. Also, it can be observed that the sensitivity of the DeFET, which is the slope of the line shown in Figure 8.4, is about $51.7\frac{\text{mV}\mu\text{m}}{\text{V}}$.

To be confident that the sensor is working properly and measuring the actual gradient in the electric field, the configuration shown in Figure 8.2 has been constructed in the Coulomb environment using the same values shown in Table 8.1. The sensor has been tested in air, and the result is shown in Figure 8.4.

The small discrepancy between the measured and the simulation results can be interpreted as follows:

- There are other coated layers associated with the CMOS 0.18-μm technology, such as the passivation layer. In the simulation only the three main layers have been used (silicon, silicon oxide, and polysilicon).
- There are some trapped charges on the CMOS layers, such as the excess charges on the silicon oxide layer, which will increase the electric field intensity above the DeFET (this was neglected in the Coulomb simulation).
- The metal wires used for biasing and conveying the currents and voltages to the output pads can also increase the intensity of the applied electric field.

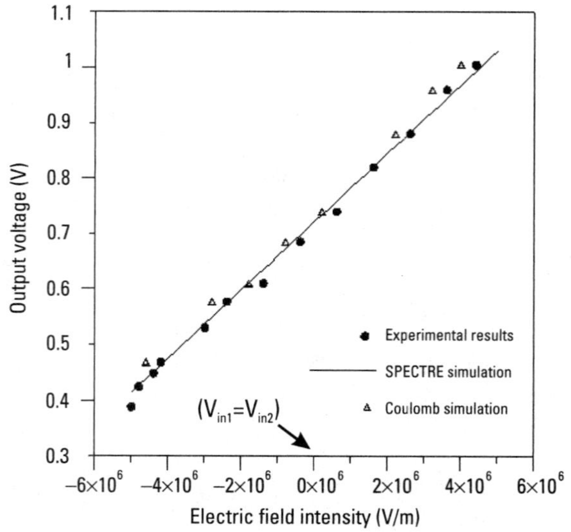

Figure 8.4 The DC response of the DeFET. (*From:* [12]. Reprinted with permission.)

Also, the DeFET is tested with a coating of silicon rubber, about 20-μm thickness. Assuming 20 μm as the thickness, we used Coulomb to compare between the experimental and simulation results. The relative permittivity of the silicon rubber is 11.7. Table 8.2 shows the measured and the simulated gradient in the electric field intensity for different electric field configurations using silicon rubber. From Table 8.2 we can confirm the operation of the DeFET.

The difference between the simulation and experimental results is due to:

Table 8.1
The Measured and Simulated Values of the DeFET Coated with Silicon Rubber

Electric Field Profile Configuration	The Measured Electric Field Intensity (V/μm)		Coulomb Results (V/μm)	
	DeFET 1	**DeFET 2**	**DeFET 1**	**DeFET 2**
Electrode 1 and 2 = +5V, electrode 4 = +5V, and electrode 3 is not connected	0.738	2.812	0.523	2.324
Electrode 1 and 2 = +5V, electrode 3 = +5V, and electrode 4 is not connected	0.323	0.572	0.283	0.347
Electrode 1 = +5V, electrode 2 = −5V, and electrodes 1 and 4 are not connected	0.317	0.627	0.192	0.451

Source: [1].

Table 8.2
The Measured Output Voltage for Different DeFET Sensors in Different Media with Different Electric Field Configurations

Electric Field Profile Configuration	The Measured Output Voltage (Peak to Peak) of DeFET 1		The Measured Output Voltage (Peak to Peak) of DeFET 2	
	Air	**Silicon Rubber**	**Air**	**Silicon Rubber**
Electrode 1 and 2 = +5V p-p, electrode 4 = +5V p-p, and electrode 3 is not connected. The frequency is 10 MHz.	0.269V	0.541V	0.216V	0.482V
Electrode 1 and 2 = +5V p-p, electrode 3 = +5V p-p, and electrode 4 is not connected. The frequency is 10 MHz.	0.069V	0.186V	0.188V	0.341V
Electrode 1 = +5V p-p, electrode 2 = –5V p-p, and electrodes 1 and 4 are not connected. The frequency is 10 MHz.	0.074V	0.175V	0.103V	0.128V
Electrode 1 = +5V p-p, electrode 3 = +5V p-p, and electrodes 2 and 4 are not connected. The frequency is 10 MHz.	0.053V	0.066V	0.345V	0.39V

Source: [1].

1. An assumption that the thickness of the silicon rubber layer is 20 μm. However, it may be thicker than this value.
2. The effect of the biasing wires, which disturb the gradient of the applied electric field.
3. The trapped charges in the different CMOS fabrication layers that disturb the electric field gradient.

8.1.2 The AC (Frequency) Response

In the AC (or frequency) response, we are interested only in the change of the output reading of the DeFET with variation of the frequency. Thus, the electric field profile was kept constant by connecting the four electrodes around the sensors to the same constant input voltage. There are two variables: the electric field intensity and the frequency of the applied electric field. To test the AC response, the frequency will be varied while keeping the electric field profile constant during the experiment. Also, to test the effect of different media on the frequency response of the DeFET, the chip has been coated with silicon rubber.

The frequency response of the DeFET in air and silicon rubber is experimentally tested and shown in Figure 8.5. In this figure, the DeFET works as a bandpass filter with bandwidths of 11 MHz and 12.5 MHz in air and silicon rubber, respectively. The quality factor ($Q = f_o/BW$) is 2.182 and 1.864 in air and silicon rubber, respectively. To confirm the measured results, the bandwidth in air has been measured using a network analyzer (HP 4395A Network/Spectrum/Impedance Analyzer) (see Figure 8.6).

Figure 8.5 shows that both the bandwidth and the central frequency depend on the type of the media that the DeFET works in. Also, it proves that the AC model proposed in Chapter 4 can represent the DeFET. This result is also important as the DeFET can be used to characterize the media above it and to extract useful information about its capacitance.

8.1.3 Other Features of the DeFET

To test the response of the DeFET with the changing of the electric field profile, the frequency was kept constant (10 MHz), and the applied electric field profile was changed by changing the magnitude, sign, and configuration of the applied voltage on the electrodes. Also, to examine the transient response of the DeFET with different electric field profiles and in different media (i.e., air and silicon

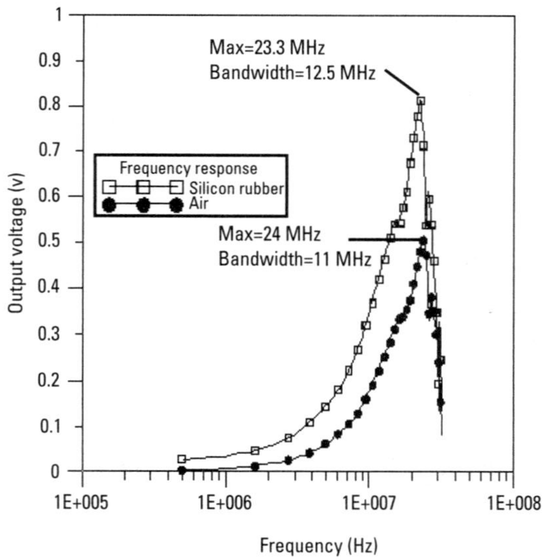

Figure 8.5 The measured frequency response of the DeFET in different media. (*From:* [12]. Reprinted with permission.)

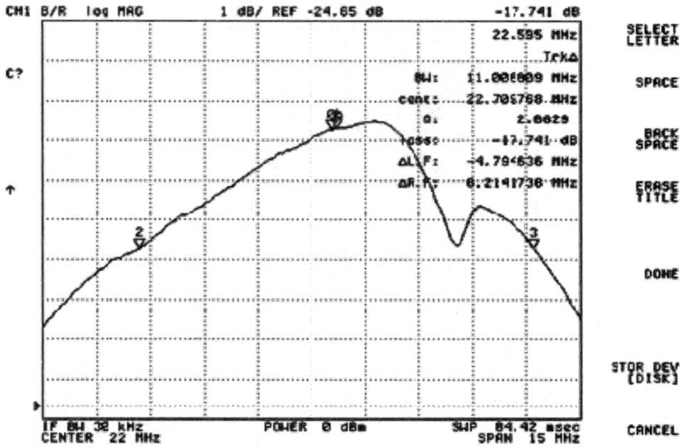

Figure 8.6 Spectrum analyzer graph shows the frequency response of the DeFET and confirms the measured values in Figure 8.5 (v_o is decibels versus frequency, and the input impedance of the spectrum analyzer is 50 ohms).

rubber), the electrodes were connected to voltage sources with different values and phases.

The results are shown in Table 8.2. From this table it can be verified that the DeFET is responding to different electric field patterns and different media, as it has been mentioned and proved before in Chapter 7 (Section 7.7). In Table 8.2, to get an out of phase (180° phase) voltage, a four-channel waveform generator (TTi TGA1240 Series Universal Waveform Generator) has been used. It has four controllable channels; each channel can operate as a fully functioning generator with controllable magnitude and phase [2].

Other dynamic and static characteristics of the DeFET have been measured. The input offset voltage is 25 μV (average value over 5 chips). The sensitivity of the DeFET is $51.7\frac{\text{mV}\mu\text{m}}{\text{V}}$. The rise and fall times have been measured using a square signal; they are 17 ns and 15 ns, respectively. The oscilloscope output taken at 100 kHz is shown in Figure 8.7. This figure illustrates the behavior of the proposed DeFET sensor when excited by triangular, square, and sinusoidal sources. The dynamic range of the DeFET is high, as mentioned before, because there is a low-noise floor. A summary of these results is shown in Table 8.3.

8.2 Noise Analysis

Noise is defined as an undesired disturbance within the frequency band of interest, or it is a disturbance that affects a signal and that may distort the information carried by the signal [3, 4]. Noise limits the minimum signal level that the DeFET sensor can process with acceptable quality.

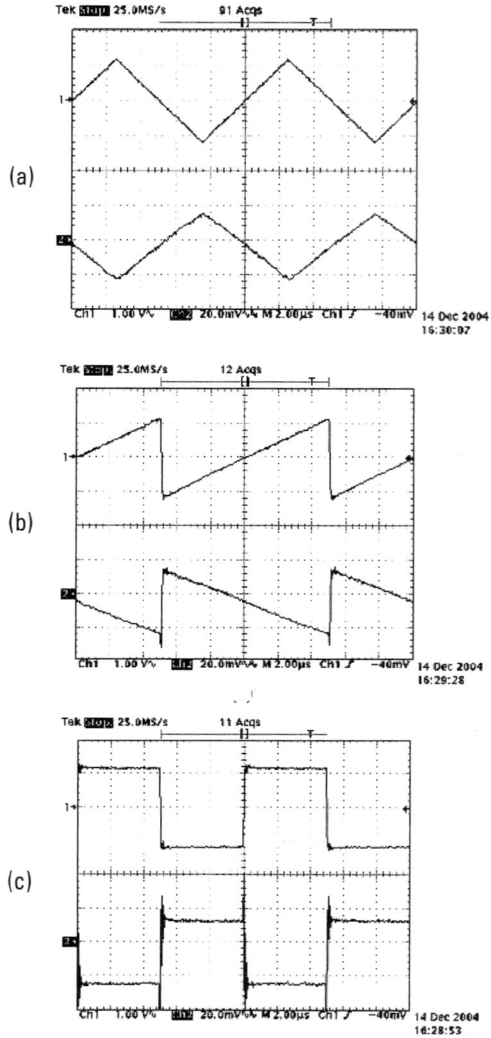

Figure 8.7 The oscilloscope showing the proposed DeFET performance at 100 kHz: (a) triangle, (b) ramp, and (c) square waveform. (*From:* [1]. Reprinted with permission.)

8.2.1 Noise Sources

There are many types of noise associated with semiconductor devices, such as shot noise, thermal noise, and flicker noise. Shot noise is defined as the noise caused by random fluctuations in the motion of charge carriers in a semiconductor, as there is always a recombination between the electrons and holes [3]. Mathematically, we can define shot noise current as follows:

$$I_n = \sqrt{2qI_{DC}BW} \tag{8.1}$$

Table 8.3
Summary of the DeFET Features (Average Values over 5 Chips)

Parameter	Value	Unit
Die area	0.0005	mm^2
Supply voltage	+/– 3.3	Volt
Sensitivity	51.7	mV μm/V
Signal-to-noise ratio	>78.2	dB
Offset voltage	25	μV
Bandwidth	Bandpass with BW = 11 MHz	Quality factor = 2.12
DC power consumption (the whole chip = 1.23 mW)	0.102	mW
Rise time	17	ns
Fall time	15	ns
Noise level	Very low	

Source: [1].

where q is the electron charge which equals 1.6×10^{-19}C, I_{DC} is the average DC current in the semiconductor, and BW is the frequency bandwidth over which the noise is considered.

Thermal noise is created as a result of the random motion of charge carriers within a conductor. In a MOSFET, the thermal noise is generated in the channel, and it can be calculated as follows [4]:

$$\overline{i_n^2} = 4KT\gamma g_m \Delta f \tag{8.2}$$

where K is the Boltzmann's constant (1.38×10^{-23} Joules/Kelvin), T is the temperature degree in Kelvin, γ is a coefficient, Δf is the bandwidth, and g_m is the transconductance of the transistor.

Flicker noise is associated with crystal surface defects in semiconductors. The noise power is proportional to the bias current, and, unlike thermal and shot noise, flicker noise decreases with frequency, so it is also called $1/f$ noise [4].

8.2.2 Noise Measurements

The noise floor of the DeFET is measured using the network analyzer (HP 4395A Network/Spectrum/Impedance Analyzer). The experimental result is plotted in Figure 8.8. From this figure, we can observe that the noise level is low, thus, the signal-to-noise ratio, and consequently the dynamic range, will be high. Also, by using cross-coupling between the p eFET and n eFET at the output, the output

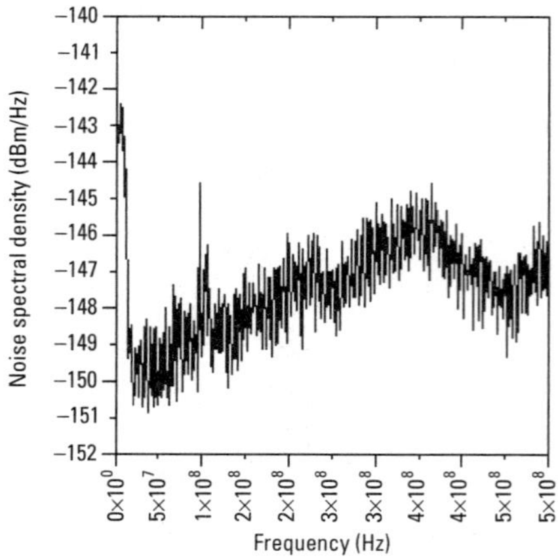

Figure 8.8 The measured noise floor of the DeFET using a spectrum analyzer. (*From:* [1]. Reprinted with permission.)

currents from the p eFET and n eFET are subtracted and any common mode noise will be rejected. Thus, a low noise floor can be obtained although floating gates are used. As an example, the signal-to-noise ratio was measured using the spectrum analyzer, as shown in Figure 8.9, using an input voltage = 5V p-p connected to the four electrodes, and frequency = 12 MHz. From Figure 8.9 we find that the dynamic range (i.e., signal-to-noise ratio in decibels) is equal to 78.

8.3 The Effect of Temperature and Light on DeFET Performance

The effect of temperature fluctuations (around room temperature) on the performance of the DeFET was studied by testing the DeFET at room temperature

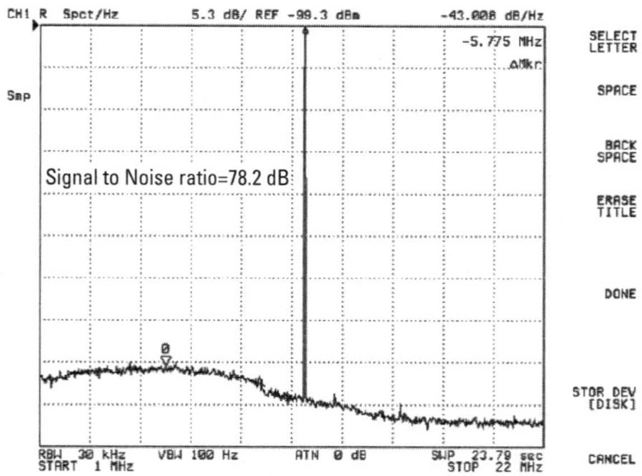

Figure 8.9 Spectrum analyzer picture shows an example of the signal-to-noise ratio (i.e., dynamic range = 78.2 dB) at 12-MHz frequency. (*From:* [1]. Reprinted with permission.)

and above it by about 10 degrees. The test results are shown in Figure 8.10; as can be seen, the temperature fluctuation has no effect on the performance of the DeFET.

To check the effect of light on the performance of the DeFET, the output voltage was measured first in the dark, then the DeFET was exposed to a light source (100-W lamp, very close to the chip), and the output voltage was remeasured. The output voltage in both cases is shown in Figure 8.11. The light has a significant effect on the output voltage as the output voltage decreased by about 32 mV when the DeFET was exposed to light. This decrease in output can be interpreted as follows.

The DeFET sensor, which consists of a p eFET and a complementary n eFET, can be considered as a CMOS image sensor at optical wavelength. The

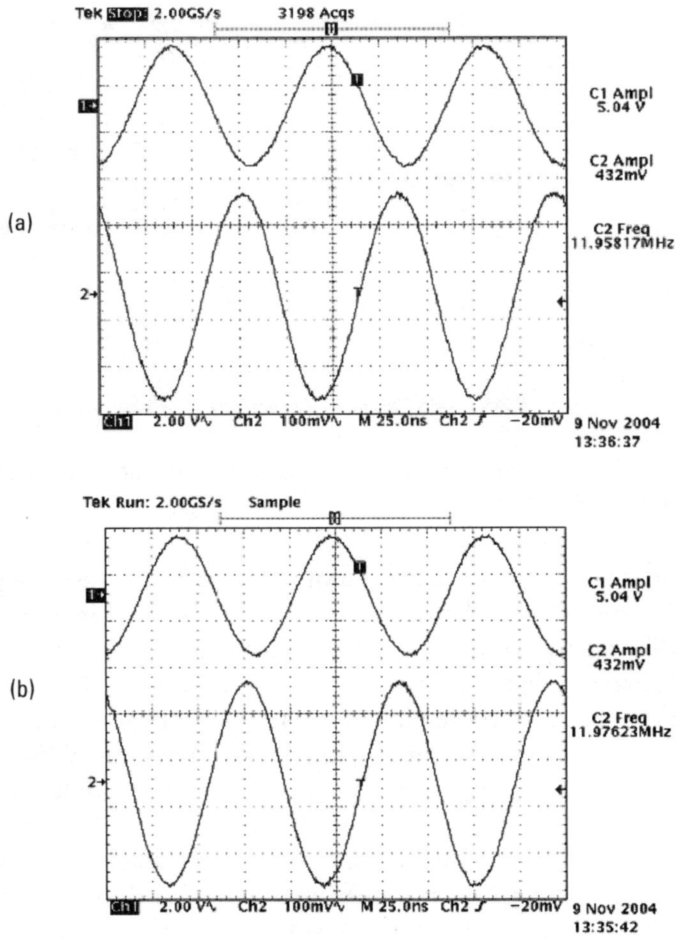

Figure 8.10 (a) The response of the DeFET at room temperature. (b) The response of the DeFET at temperature 10 degrees above the room temperature.

(a)

(b)

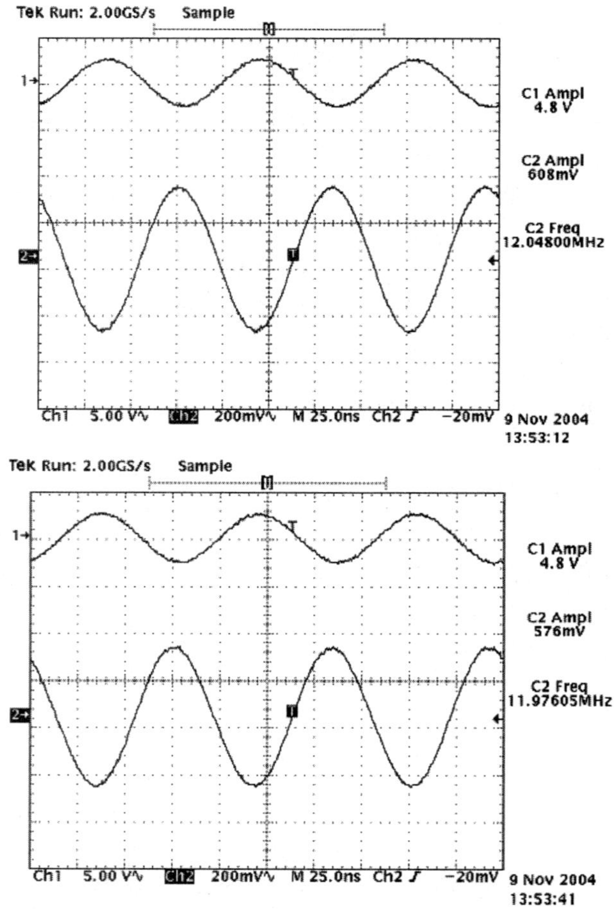

Figure 8.11 (a) The response of the DeFET with no illumination. (b) The response of the DeFET with a very close 100W-light source.

cross-sectional view of the p eFET is shown in Figure 8.12. When the p eFET is illuminated, the incoming photons generate electron-hole pairs in the channel of the p eFET, the well-to-drain and source depletion region, and the well-to-substrate depletion region (see Figure 8.12). They are separated vertically by the intrinsic vertical electric field resulting from the vertical band structure, caused by the difference in work function between the gate and channel [5]. The holes move toward the minimum potential point in the channel while the electrons move in the opposite direction (i.e., away to the bottom of the depletion region). For electron/hole pairs generated in the n-well/source (drain) depletion region, the built-in electric field separates them in the vertical direction resulting in holes being collected in the channel. The number of excess holes collected by the channel is determined by the optical spectrum of the light as well as other device parameters. Thus, finally there are an excess number of carriers (holes)

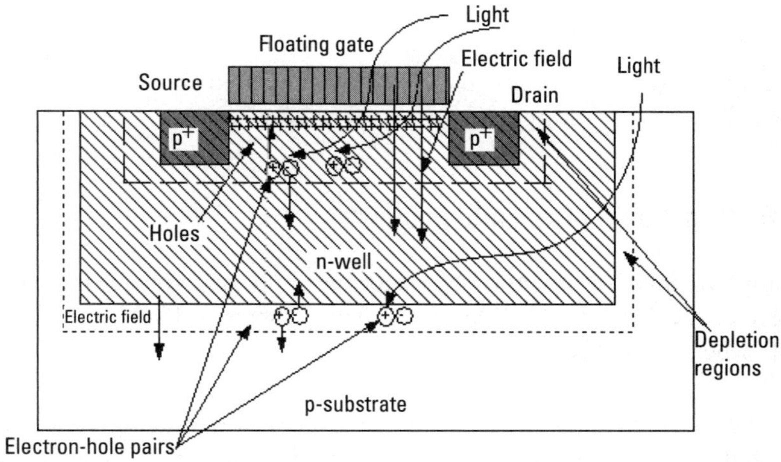

Figure 8.12 Cross-section view of the p eFET.

in the channel, which come from the channel itself, the well-to-source (drain) depletion region, and the well-to-substrate depletion region; consequently, the drain-to-source current will be increased.

Similarly, for the n eFET the same situation exists. However, in the n eFET the n-well/p-substrate does not exist so the number of electrons in the channel will be increased but not by the same ratio as of the p eFET. Therefore, the output current is the difference between n eFET and p eFET ($I_n - I_p$), and as this difference is decreased the output current and consequently the output voltage will be decreased, as can be seen in Figure 8.11.

Under illumination, the output current can be calculated as:

$$i_{out} = i_{ele} - i_{opt} \tag{8.3}$$

where i_{out} is the total output current of the DeFET, I_{ele} is the output current due to the electric field effect, and I_{opt} is the output current due to the light effect.

The optical current under illumination is proportional to the optical power p, as follows [5, 6]:

$$i_{opt} = \alpha P \tag{8.4}$$

where α is a constant that depends on the device technology and the generation rate of electron/hole pairs. In the present case $\alpha = \alpha_p - \alpha_n$, where α_p and α_n are the p eFET and n eFET constants.

If the electric field is kept constant, then i_{ele} in (8.3) will be constant, and i_{out} will be linearly related to the optical power and consequently to the illumination intensity that induces this current, as shown in (8.5):

$$i_{out} = \text{constant} - \alpha P \tag{8.5}$$

Equation (8.5) shows that the DeFET can be used as a visual imager. This will be a direction for future work.

8.4 Testing the Electric Field Imager

A microphotograph of the proposed electric field imager, which is implemented in target TSMC 0.18 μm CMOS technology, is shown in Figure 8.13. As mentioned earlier, it consists of two main parts:

- The actuation part, which is a quadrupole electrode configuration to generate a controllable nonuniform electric field profile, and consequently a DEP force to levitate the cells;
- The sensing part, which is an array of the DeFET [7]. The size of each electrode is 100 μm from edge-to-edge. The size of the DeFET is 25 μm $\times$ 20 μm.

8.4.1 The Response of the Imager Under Different Environments

To test the response of the imager under different environments, the chip was coated with silicon rubber. Then the AC response of the imager was experimentally tested with different electric field profiles. Figures 8.14 and 8.15 show the measured output current for each DeFET at a high frequency (10 MHz) and using two different media (i.e., air and silicon rubber). In Figure 8.13, Electrodes 1 and 3 are connected to –5V p-p, and electrodes 2 and 4 are connected to +5V p-p (Q = quadrupole configuration). The result is shown in Figure 8.14, while Figure 8.15 shows the result of another electrode configuration [i.e., electrodes 1 and 4 are connected to 5V peak-to-peak (p-p), electrode 2 is connected to –5V p-p, and electrode 3 is not connected].

Figures 8.14 and 8.15 show the measured voltage, which varies with the changing of the media and the applied electric field profile. Thus, it can be confirmed that the imager is working properly with different media.

8.4.2 Testing the Imager with Biocells

To experimentally test that the effect of the biocells on the applied electric field can simultaneously be actuated and sensed, the imager has been tested with polystyrene microspheres with a relative permittivity of 2.5 and diameters of 8.9 μm, and 20.9 μm, respectively. The particles are suspended in deionized water with a

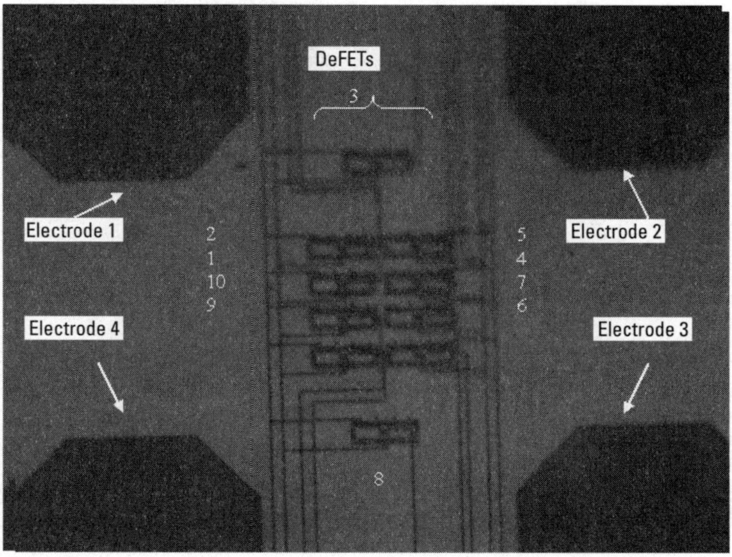

Figure 8.13 The die picture shows the quadruple electrodes, and the DeFET sensors are in between.

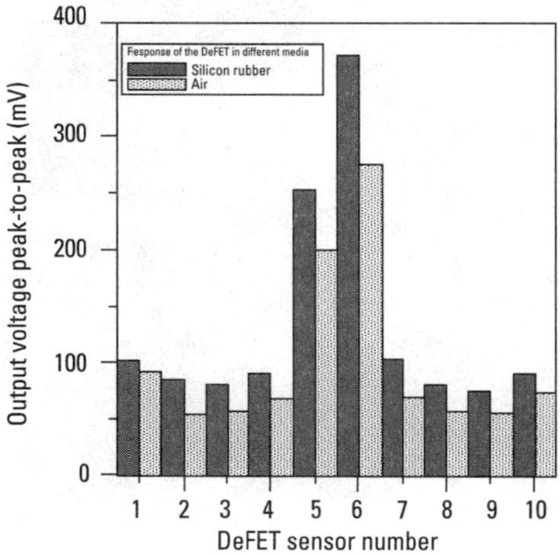

Figure 8.14 The measured output voltage for different DeFET sensors with the configuration of: electrodes 1 and 3 = –5 p-p, and electrodes 2 and 4 = +5V p-p (i.e., quadrupole configuration). The frequency is 10 MHz.

measured conductivity between 1.3–1.9 μS/cm. A reflectance microscope is used to observe the particles from the top. The arrangement of the various laboratory components used during experimentation is shown in Figure 8.16.

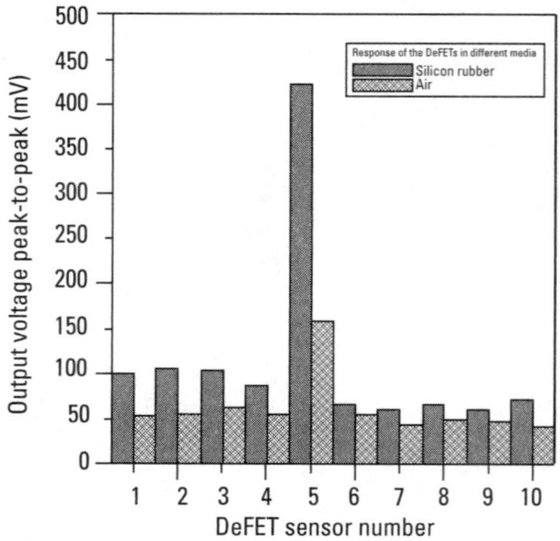

Figure 8.15 The measured output voltage for different DeFET sensors with the configuration of: electrodes 1 and 4 = 5 p-p, electrode 2 = –5V p-p, and electrode 3 is not connected. The frequency is 10 MHz.

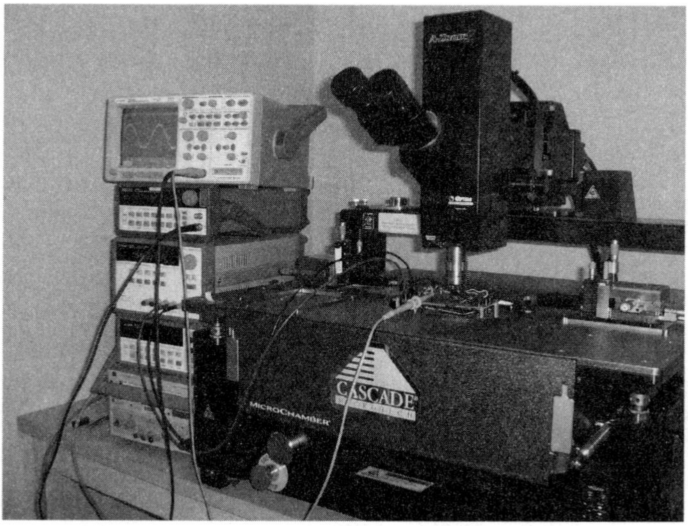

Figure 8.16 The equipment used for testing the imager with polystyrene microsphere cells.

To contain both the suspending medium and particles under test, a fluidic chamber of approximately 10 μL was created, using a similar idea to that proposed in [8]. The electrodes and the DeFET sensors form the bottom of the chamber and silicon rubber forms the walls of the chamber. A small piece of glass covers the chamber.

A quadrupole configuration is needed to trap and levitate a biocell at the central point of the imager. To achieve this configuration, electrodes 2 and 4 are connected to a +5V peak-to-peak sinusoidal signal, and electrodes 1 and 3 to −5V peak-to-peak (i.e., 180 degrees out of phase, sinusoidal signal, with a frequency of 3 MHz). Figures 8.17 and 8.18 show a levitated cell, after the electric field is turned on, just above the central point of the imager. The levitated cell leaves the focal range of the microscope and appears out of focus.

Before adding the fluid, the readings of the DeFET sensors were taken as references, and then during the levitation, the readings of the DeFET sensors were taken again. This allows for simultaneous actuation and sensing (measure)

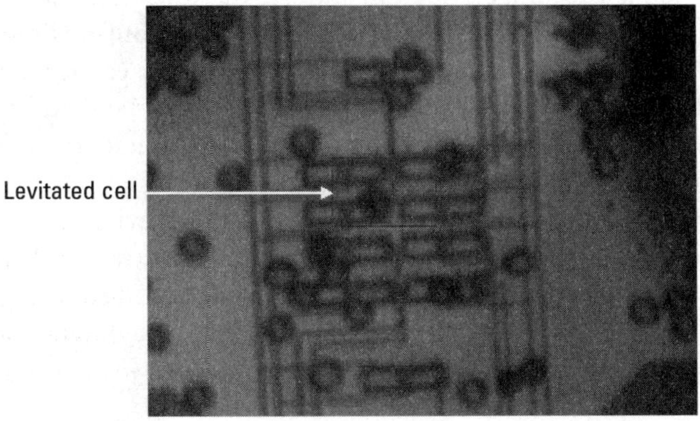

Figure 8.17 Levitated polystyrene cell with diameter 8.9 μm. (*From:* [9]. Reprinted with permission.)

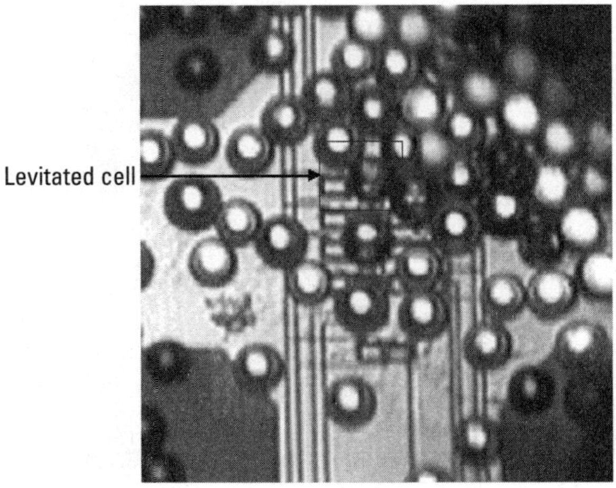

Figure 8.18 Levitated polystyrene cell with diameter 20.9 μm. (*From:* [9]. Reprinted with permission.)

of the effect of the cells on the electric field intensity profile. The results confirmed the simulation results that are shown before in Chapter 8.

Figure 8.19 shows the measured output voltage of the DeFET sensors numbers 1, 4, 7, and 10 respectively, in air (without fluid), and after adding the suspending fluid, which contains 8.9-μm, and 20.9-μm diameter polystyrene microspheres cells [9]. From this figure, it can be observed that DeFETs 1 and 10 show higher readings compared with DeFETs 4 and 7. When comparing this result with the picture in Figure 8.17, it can be observed that there are two cells above DeFETs 1 and 10, while there are no cells above DeFETs 4 and 7, so the readings of them are lower than 1 and 10. Also, for the cells with 20.9-μm diameters, DeFETs 1, 4, and 10 show high readings, while DeFET 7 does not show the same high reading, so we can surmise that there are cells above them, while there are no cells above DeFET 7. Figure 8.18 confirms these observations. Another important measurement is that the output voltage increases with increasing diameter of the biocells. Thus, the imager can be used in biomedical applications not only for cell detection, but also to differentiate between the diameters of different cells based on its output reading.

This result is very important when trapping and levitating only one cell, and then an electric field image about the cell can be extracted. If this is done with many types of cells then we can get what can be called a fingerprint for each one of these cells, which could be a new technique to differentiate between the cells using their electrical properties instead of their visual image, which is unreliable.

To verify that a real time electric field image can be obtained during the manipulation of the biocells under the effect of the electric field, the real time readings of DeFETs 1, 4, 7, and 10 have been taken at different sample times.

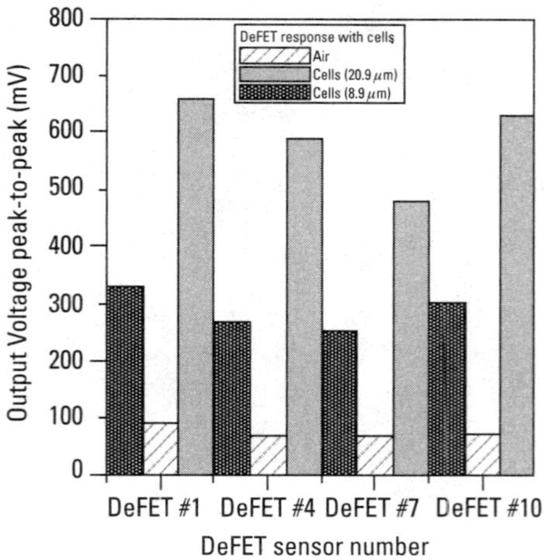

Figure 8.19 The DeFET sensor's response in air and in fluid contains different cell sizes.

Figure 8.20 shows the polystyrene microspheres with an 8.9-μm diameter suspended in deionized water with a measured conductivity between 1.3–1.9 μS/cm. At each time shown in this figure, the readings of the four DeFETs just around the central point have been recorded [i.e., DeFET number 1, 4, 7, and 10 (see Figure 8.13)].

From Figure 8.20, it can be directly determined if there is a cell above the DeFET sensor or not, by just reading its output voltage. It can be noticed that the output readings are not the same in each case, because the cells are not the same height above the DeFET sensors, so the DeFET can be used as a levitation height sensor in DEP applications. This result will be a new direction for future research.

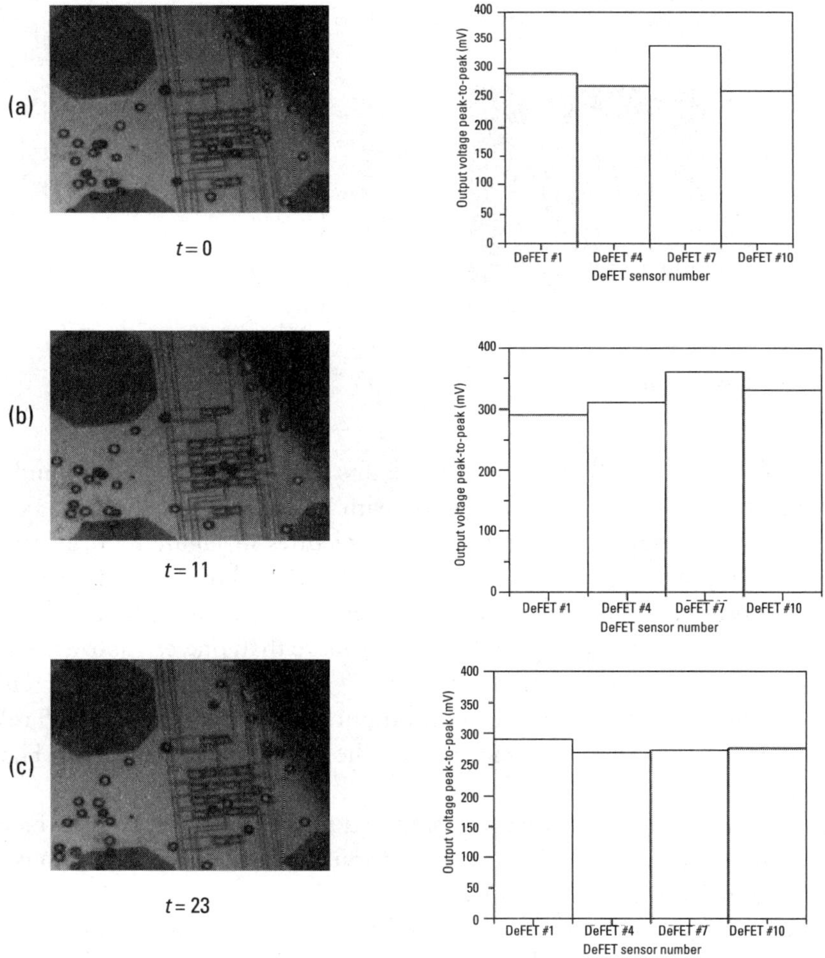

Figure 8.20 (a–f) Different sample times and the real-time readout from the DeFET sensors.

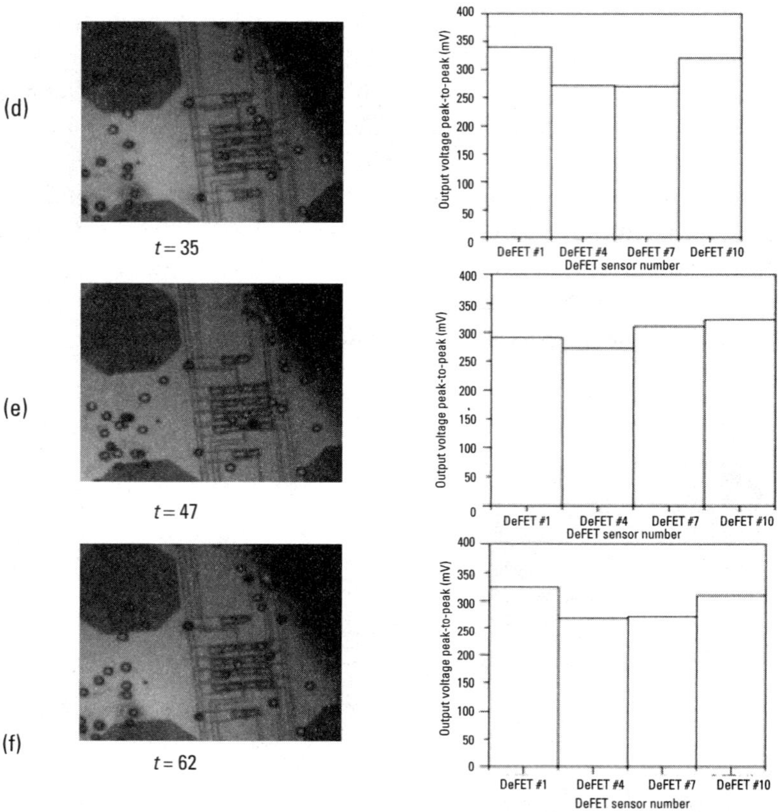

Figure 8.20 (continued)

Table 8.4 summarizes all of these results. One of the important and interesting observations in this table occurs with sensor number 7 at $t = 11$s and $t = 0$s. By comparing these results with the pictures in Figure 8.20, it can be observed that sensor number 7 at $t = 11$s has two cells above it and it reads 360 mV at the output. While at $t = 0$s, it reads 340 mV, so we have about a 20-mV difference in the output voltage when there is more than one cell above the sensor. Also, sensor number 1 at $t = 35$s and $t = 62$s, has the same situation as sensor number 7; however, the difference in the output reading is lower (about 15 mV). This is because the cell is not exactly above the sensor, as can be seen in Figure 8.20(d, f).

Previous results show that the proposed sensor can be used as a biocell counter or as a detector of the biocell location. Both of these applications are very important in the biomedical field [10–12].

From Table 8.4, it can be confirmed that the DeFET is responding with the change of the electric field intensity and not with the change of the light illumination, for the following reasons:

Table 8.4
Summary of the Real-Time Readings in mV at Different Time Snapshots

DeFET Number	Air	$t = 0s$	$t = 11s$	$t = 23s$	$t = 35s$	$t = 47s$	$t = 62s$
1	92	290	290	290	340	291	325
4	69	269	310	269	271	272	268
7	70	340	360	273	270	310	271
10	71	260	330	276	320	322	310

- The response of the DeFET sensor was measured under different environments with different electric field profiles without illumination (Section 8.2.1) and it was confirmed that the sensor is responding very well to these changes.
- The response of the DeFET sensor was tested under different input waveforms without illumination [i.e., sinusoidal, triangle, and rectangle waveforms (Section 8.1.3), and the sensor shows a good response with these different waveforms.
- The maximum light effect that was measured before in Section 8.3, gives about a 32-mV change in the output voltage. However, there is more than a 200-mV change in the output voltage under the effect of the electric field. For example in Table 8.4, for DeFET number 7 at $t = 11s$ (a cell above it) the reading is 360 mV. At $t = 35s$ (no cells above it) the reading is 270 mV, so the difference due to the position of the cell is about 90 mV. If it is assumed that there is a 32-mV change due to illumination (worst case scenario), then the remaining (i.e., 60 mV) is due to the electric field. Thus, the electric field measurement dominates the reading.

Based on all these reasons it is certain that the sensor's reading variation is due to the electric field and not due to the high illumination.

8.5 Packaging the Lab-on-a-Chip

To simultaneously monitor the levitated cells during the measurement of the output voltage from the imager sensors, a horizontal viewing architecture was needed to monitor the vertical displacement of the cells during measurement. To achieve that, a similar technique to that proposed by Hartely et al. [13] was used. This technique was used successfully to trap and monitor a single cell [13]. The assembly of the carrier system is described next.

A glass spacer (3-mm thick) was bonded into the chip package (40 PIN DIP IC carrier) well using Devcon adhesive glue. The imager chiplet (Figure 8.1) was subsequently bonded on top of the spacer, as shown in Figure 8.21. Electrical connections (wire bonds) were made from the chiplet pads to the chip package bond pads. Finally, the chamber was mounted above the chiplet. The chamber housing itself, shown in Figure 8.22, was machined out of Lucite plastic fitted with horizontal view ports.

To facilitate the particle injection, a glass microtubule (1-mm outside diameter, ~120-μm inside diameter) was bonded into the vertical shaft drilled into the upper surface of the chamber cavity. The Lucite block was then carefully positioned and aligned, such that the microtubule tips were centered along the vertical axis of the quadrupole electrodes and positioned approximately 600-μm above the chiplet surface. The Lucite block, once appropriately positioned, was secured to the chip packager using Devcon glue. To complete the chamber assembly, 4-mm long needles were glued (epoxy) onto the chambers primary fluid ports, as shown in Figure 8.22. Figure 8.23 shows a complete system that includes the elevator (i.e., x, y, z table), which raises the system.

Figure 8.21 The glass spacer.

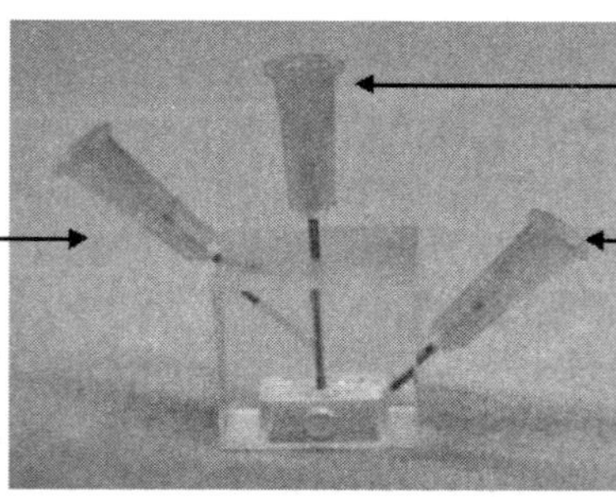

Figure 8.22 The fluidic chamber.

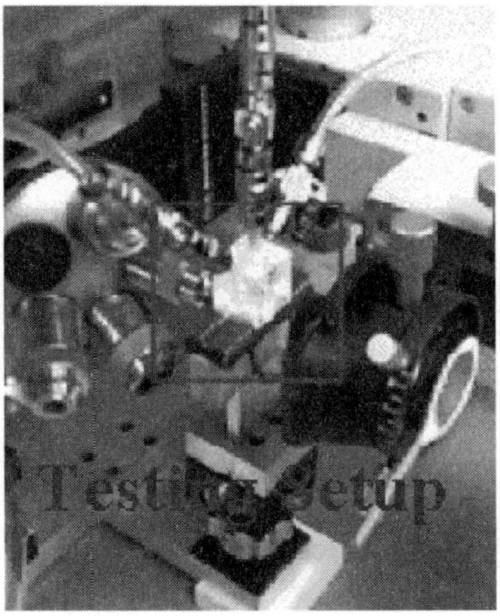

Figure 8.23 A complete system.

References

[1] Ghallab, Y. H., and W. Badawy, "DeFET, a Novel Electric Field Sensor for Lab-on-a-Chip and Biomedical Applications," *IEEE Journal of Sensor,* Vol. 6, No. 4, August 2006, pp. 1027–1037.

[2] Sze, S., *Physics of Semiconductor Devices,* 2nd ed., New York: McGraw-Hill, 1999.

[3] Abidi, A., "High Frequency Noise Measurements on FETs with Small Dimensions," *IEEE Trans. on Electronic Devices,* Vol. 33, 1986, pp. 1801–1805.

[4] Razavi, B., Chapter 7, in *Design of Analog CMOS Integrated Circuits,* New York: McGraw-Hill, 2001.

[5] Zhang, W., and M. Chan, "A High Gain N-Well/Gate Tied PMOSFET Image Sensor Fabricated from a Standard CMOS Process," *IEEE Trans. on Electron Devices,* Vol. 48, No. 6, 2001, pp. 1097–1102.

[6] Yang, E. S., *Microelectronic Devices,* 2nd ed., New York: McGraw-Hill, 1988.

[7] Ghallab, Y. H., and W. Badawy, "Magnetic Field Imaging Detection Apparatus," U.S. Patent, 7,405,562, July 2008.

[8] Keilman, J. R., G. A. Jullien, and K. V. I. S. Kaler, "A Programmable ac Electrokinetic Micro-Particle Analysis System," *IEEE Proceedings of Biomedical Circuits and Systems (BioCAS04),* 2004, pp. S2.3-9–S2.3-12.

[9] Ghallab, Y. H., and W. Badawy, "A Single CMOS Chip for Biocell Trapping, Levitation, Detection and Characterization," *International Symposium on Circuit and System (ISCAS 06),* Greece, May 2006, pp. 3349–3352.

[10] Manaresi, N., et al., "A CMOC Chip for Individual Manipulation and Detection," *IEEE International Solid-State Circuits Conference (ISSCC 03)*, 2003, pp. 486–488.

[11] Medoro, G., et al., "CMOS-Only Sensors and Manipulation for Microorganisms," *Proceedings of IEDM,* 2000, pp. 415–418.

[12] Medoro, G., et al., "A Lab-on-a-Chip for Cell Detection and Manipulation," *IEEE Sensors Journal*, Vol. 3, No. 3, 2003, pp. 317–325.

[13] Hartley, L. E., K. V. I. S. Kaler, and R. Paul, "Quadrupole Levitation of Microscopic Dielectric Particles," *Journal of Electrostatics*, Vol. 46, 1999, pp. 233–246.

9

Readout Circuits for Lab-on-a-Chip

Readout circuits play very important rules in the lab-on-a-chip integration. These circuits perform the following two main functions: amplify the weak output signal of the lab-on-a-chip sensors, and condition the output signal and consequently improve the signal to noise ratio (S/N) and the dynamic range of the output signal of the lab-on-a-chip. The availability of the readout circuits to be integrated on the lab-on-a-chip is an important advantage. This feature is reflected on the portability of the lab-on-a-chip.

In this chapter, an introduction about current-mode circuits, which are suitable to be integrated on lab-on-a-chip, will be introduced. Some current-mode readout circuits will be introduced and discussed.

9.1 Current-Mode Circuits

A current-mode circuit can be defined as a circuit in which current is used as the active variable instead of voltage either throughout the whole circuit or in certain critical circuit areas. The advantages of the current-mode circuits can be summarized as follows [1]:

- Wider dynamic range;
- Higher signal bandwidth;
- Greater linearity;
- Higher usable gain.

Analog integrated circuit design is receiving a great boast from the development and application of current-mode processing, which is rapidly superseding approaches based on the voltage-mode approach [2]. Figure 9.1 shows different building blocks for current-mode circuits.

9.2 Operational Floating Current Conveyor (OFCC)

Voltage-mode circuits using conventional operational amplifiers (op-amp) have the disadvantages of severely reduced bandwidth at higher gains, because of the op-amp's fixed gain/bandwidth product. Attempts have been made to overcome the gain-dependent bandwidth limitation, and this in turn has led to the renewed interest in circuits, which use current as their active parameter. Such attempts have led to the development of the second-generation current conveyor (CCII) [3–5], current-feedback op-amp (CFB op-amp) [6], and an operational floating conveyor (OFC) [7].

The CCII, in general, is built around a conventional op-amp [8, 9] with a current output property, and is designed to be used in open-loop. The CCII, since its introduction, has gained wide acceptance as an extremely versatile building block [10, 11]. The CFB op-amp has a voltage output property and is designed to be used in closed-loop with current feedback from the voltage output

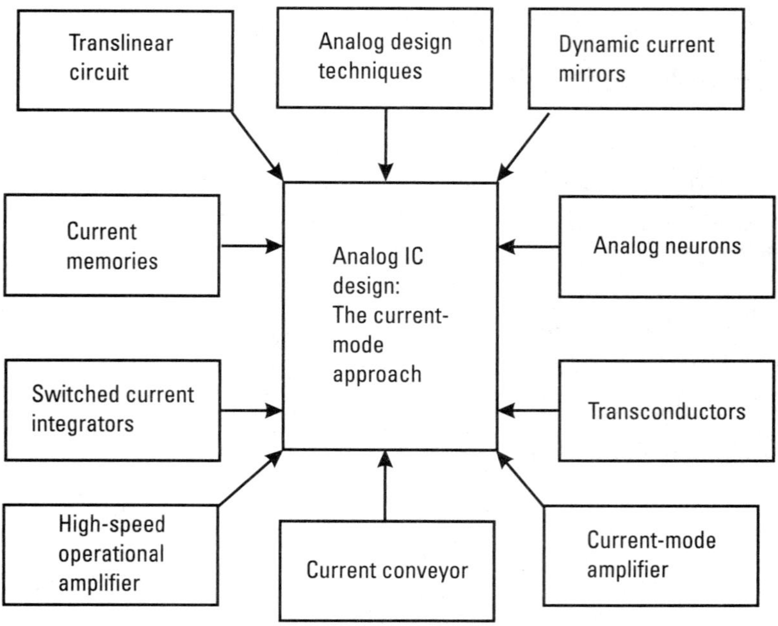

Figure 9.1 Current-mode analog building blocks.

node to the current input node. The OFC, on the other hand, is built around a CFB op-amp and combines both the features of the CFB op-amp and CCII with an additional current output. It is designed to be used in closed-loop with current feedback from the voltage output node to current input node.

All of these circuits employ current processing techniques to improve dynamic speed capability with their inherent ability to provide a constant bandwidth, independent of gain. A modification in the OFC circuit topology is obtained in order to yield a five-port general purpose analog building block, termed as an operational floating current conveyor [9–16]. The OFCC has transmission properties similar to the OFC but with an additional current output, which allows more maneuvering of the device capabilities.

The OFCC is a five-port network, comprised of two inputs and three output ports, as shown in Figure 9.2. In this diagram, the port labeled *X* represents a low-impedance current input, port *Y* is a high-impedance input voltage, *W* is a low-impedance output voltage, and *Z*+ and *Z*− are the high-impedance current outputs with opposite polarities.

The OFCC operates where the input current at port *X* is multiplied by the open-loop transimpedance gain Z_t to produce an output voltage at port *W*. The input voltage at port *Y* appears at port *X* and thus a voltage tracking property exists at the input port. Output current flowing at port *W* is conveyed in phase to port *Z*+ and out of phase with that flowing into port *Z*−, so in this case a current tracking action exists at the output port. Thus, the transmission properties of the ideal OFCC can be conveniently described as:

$$\begin{bmatrix} i_Y \\ v_X \\ v_W \\ i_{Z+} \\ i_{Z-} \end{bmatrix} = \begin{bmatrix} 0 & 0 & 0 & 0 & 0 \\ 1 & 0 & 0 & 0 & 0 \\ 0 & Z_t & 0 & 0 & 0 \\ 0 & 0 & 1 & 0 & 0 \\ 0 & 0 & -1 & 0 & 0 \end{bmatrix} \begin{bmatrix} v_Y \\ i_X \\ i_W \\ v_{Z+} \\ v_{Z-} \end{bmatrix} \tag{9.1}$$

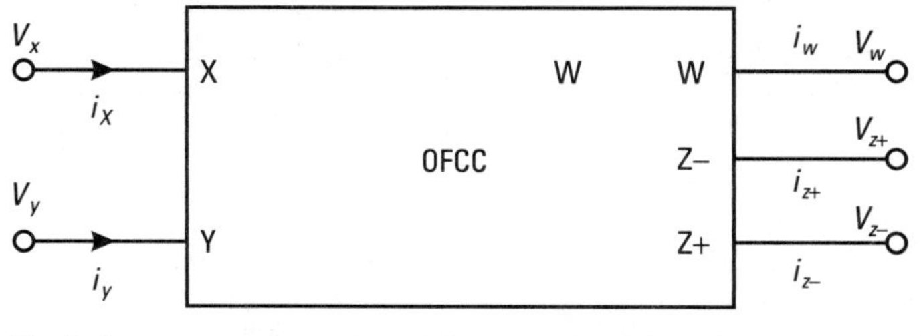

Figure 9.2 Block diagram representation of the operational floating current conveyor. (*From:* [9]. Reprinted with permission.)

where i_Y and v_Y are the inward current and voltage at the Y port, respectively. i_X and v_X are the input current and voltage at the X port, respectively. i_W and v_W are the output current and voltage at the W port, respectively. i_{Z+} and v_{Z+} are the output current and voltage at the $Z+$ port, respectively. Similarly, i_{Z-} and v_{Z-} are the output current and voltage at the $Z-$ port, respectively. Z_t represents the impedance between the X and W ports.

The OFCC can be implemented by applying the principle of supply current sensing to a current feedback (CFB) op-amp [17] such as illustrated in Figure 9.3. The current mirrors CM1 and CM2 establish the output current at port $Z+$. Also, CM1 and CM2 and their crosscoupling with the current mirrors CM3 and CM4 through the current steering transistors CS1 and CS2 generate a complementary output current at port $Z-$. The OFCC is basically designed to be used in a closed loop configuration with the current being fed back from port W to port X.

9.2.1 A Simple Model

A simple model of the OFCC based on the circuit topology shown in Figure 9.3 is illustrated in Figure 9.4. In this figure, R_X and R_Y are the resistances of the CFB op-amp at negative (−) and positive (+) ports, respectively. C_X and C_Y are the input capacitances of the CFB op-amp (−) and (+) ports, respectively. R_T and C_T are the small signal resistance and internal compensation capacitance of the CFB op-amp. R_{Z+} and R_{Z-} are the small signal output resistances of the respective current mirrors at node $Z+$ and $Z-$, respectively, at the DC operating point. C_{Z+} and C_{Z-} represent the output capacitances of the respective current mirrors at nodes $Z+$ and $Z-$, respectively.

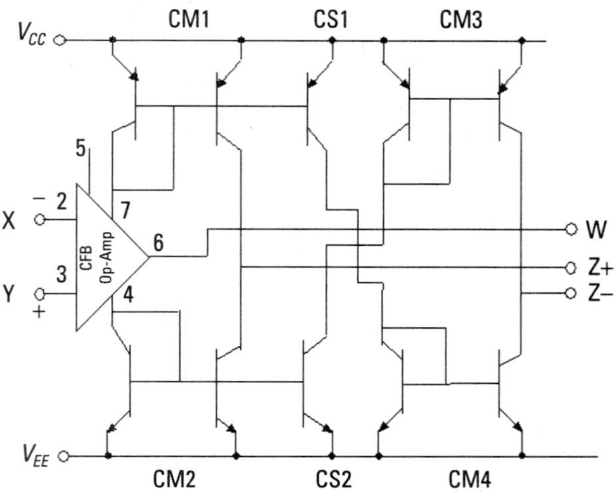

Figure 9.3 Circuit scheme of the OFCC. (*From:* [9]. Reprinted with permission.)

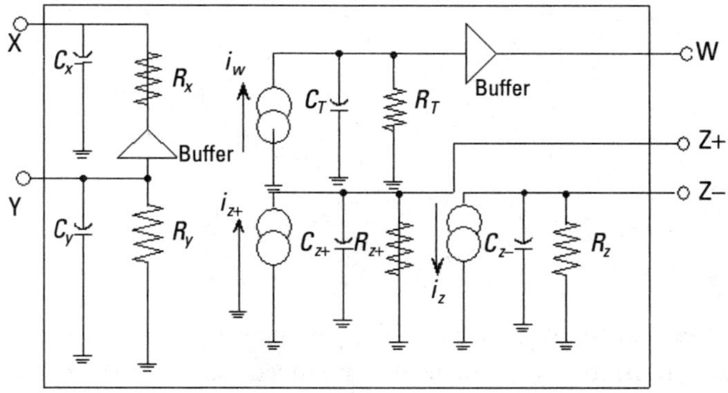

Figure 9.4 The OFCC's model. (*From:* [9]. Reprinted with permission.)

9.2.2 OFCC with Feedback

As mentioned earlier, unlike the CCII, the OFCC is designed with a feedback resistor between *W* and *X* (i.e., negative feedback between *W* and *X*). This feedback resistor allows the OFCC to operate as a positive or negative current conveyor while simultaneously reducing the input resistance at the *X* port [9]. Also, the negative feedback improves the DC stability as well as the transfer function accuracy [3, 18–20].

To understand why the input resistance at the *X* terminal is reduced by negative feedback, consider the OFCC with feedback resistor R_W, as shown in Figure 9.5. The capacitive reactance to ground due to C_X is quite high and can therefore be ignored in the frequency range of interest. In this case, the input current is defined as:

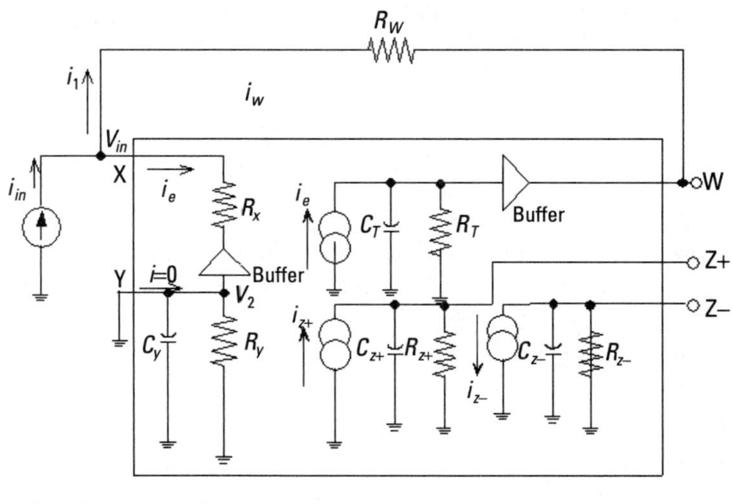

Figure 9.5 Circuit for measuring R_X.

$$i_{in} = i_1 + i_e \tag{9.2}$$

and

$$i_1 = \frac{v_{in} - v_W}{R_W} \tag{9.3}$$

where v_{in} is the input voltage at port X, v_W is the output voltage at port W, R_W is the feedback resistance, i_1 is the feedback current, and i_e is the error current.

Using (9.4), the output signal voltage v_W is given as

$$v_W = -i_e Z_T \tag{9.4}$$

where $Z_T = R_T // (1/j\omega C_T)$. For low frequencies ($\omega << 1/R_T C_T$) the reactance due to C_T is very high and can be ignored. Thus, $Z_t \approx R_T$ and

$$v_W = -i_e R_T \tag{9.5}$$

Since the Y port is connected to ground, $v_2 = 0$, and therefore the error current is defined as

$$i_e = \frac{v_{in} - 0}{R_X} \tag{9.6}$$

By substituting (9.3), (9.5), and (9.6) into (9.2), it can be expressed as

$$i_{in} = \frac{v_{in}\left(R_X + R_T + R_W\right)}{R_X R_W} \tag{9.7}$$

The low frequency input resistance at the X port can therefore be expressed as

$$R_{in} = \frac{v_{in}}{i_{in}} = \frac{R_X R_W}{R_X + R_T + R_W} \tag{9.8}$$

with typical resistor values of $R_X = 50\Omega$, $R_W = 1$ KΩ, and $R_T = 200$ MΩ, (9.8) yields $R_{in} = 0.0025\Omega$. Therefore, the input resistance at X is greatly reduced, thus minimizing the voltage tracking error between X and Y, and therefore can be neglected at low frequencies.

9.3 Current-Mode Instrumentation Amplifier

Instrumentation amplifiers are used in many application areas, such as medical instrumentation, readout circuits of biosensors, signal processing, and data acquisition [21]. Lab-on-a-chip devices, which require the integration of the actuation and sensing parts, as well as the readout circuit in a single chip, also require wide bandwidth instrumentation amplifiers to suppress any unwanted common mode signals [22]. Also, in lab-on-a-chip applications, an instrumentation amplifier has been used as a read out circuit to amplify the output signal from the lab-on-a-chip building blocks and to improve the accuracy of the measurement.

The conventional voltage-mode instrumentation amplifier (VMIA) based on a voltage operational amplifier (shown in Figure 9.6) exhibits a narrow bandwidth that is highly dependent on the gain due to the fixed gain bandwidth product of the operational amplifiers. Moreover, VMIAs require precise resistor matching to achieve a high common-mode rejection ratio (CMRR) [23–25].

Current-mode instrumentation amplifiers (CMIA) are superior to the conventional VMIA in performance in terms of CMRR and frequency range of operation, and furthermore do not require matched resistors [26–31]. Currently, most of the CMIAs use the second-generation current conveyor (CCII) [7, 8].

9.3.1 Current-Mode Instrumentation Amplifier (CMIA) Based on CCII

Figure 9.7 shows the basic CMIA in block diagram form, which uses two positive second-generation current conveyors (CCII+) [29]. The frequency-dependent differential gain [$A_d(s)$] can be computed as follows:

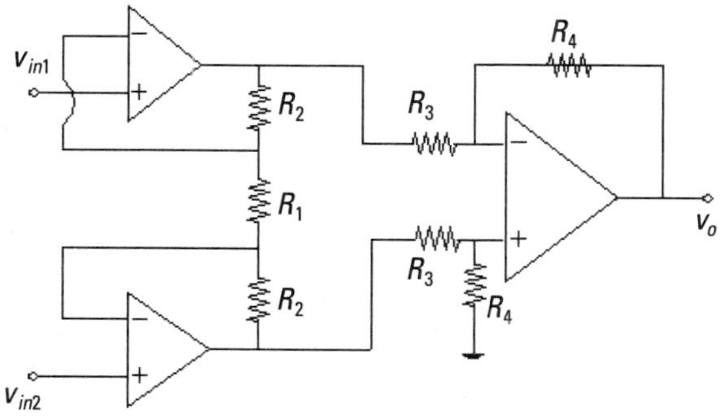

Figure 9.6 Voltage-mode instrumentation amplifier.

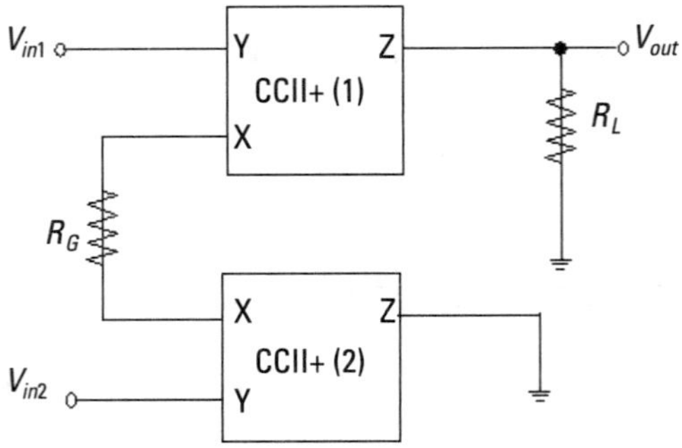

Figure 9.7 Two CCII+ CMIA [29].

$$A_d(s) = \frac{v_o}{v_{in1} - v_{in2}} = \frac{R_L}{R_G + 2R_X} \cdot \frac{1}{1 + sCR_L} \tag{9.9}$$

where R_X is the equivalent input resistance at the X terminal, R_G is the gain setting resistor, C is the effective output capacitance of the CCII+, and R_L is the load resistor. The two main advantages of using the two CCII+ topologies are that it presents a simple and symmetric circuit topology, and it has a high CMRR without requiring matched resistors.

Another CMIA design approach, as shown in Figure 9.8, uses three CCII+ [30]. In this case, the differential gain $A_d(s)$ is given as follows:

$$A_d(s) = \frac{v_o}{v_{in1} - v_{in2}} = \frac{2R_L}{R_G + 2R_X} \cdot \frac{1}{1 + sCR_L} \tag{9.10}$$

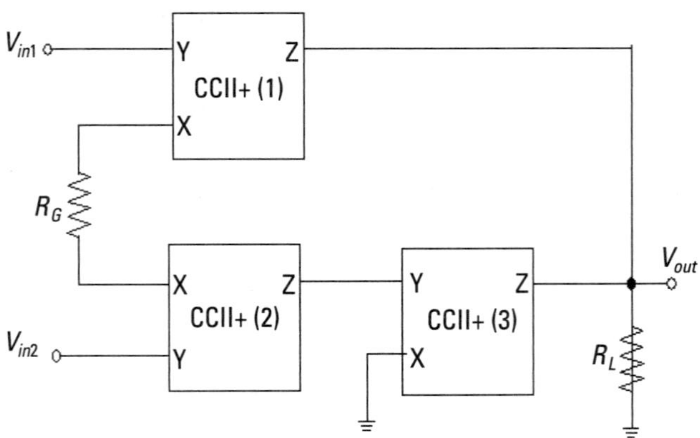

Figure 9.8 Three CCII+ CMIA. (*From:* [26]. Reprinted with permission.)

This design approach yields a higher CMRR and higher differential gain compared to the two CCII+ approach. In both of the above cases, the differential gain is inversely proportional to R_X [see (9.9) and (9.10)] and the accuracy is limited by the tolerance of R_X, which is low.

To improve the accuracy of the CMIA [3], two op-amps working in conjunction with two CCII+ are used as shown in Figure 9.9. The differential gain of this CMIA can be shown to be:

$$A_d(s) = \frac{v_o}{v_{in1} - v_{in2}} = \frac{R_L}{R_G} \cdot \frac{1}{1 + \frac{s\tau}{1 + K\beta}} \tag{9.11}$$

where $\beta = \frac{R_G}{2R_X + R_G}$, R_L is the load resistance, and τ and K are the time constant and the low frequency gain of the op-amp, respectively. The low frequency gain, in this case, is independent of R_X [see (9.11)]. However, the bandwidth is still dependent on R_X. Moreover, this approach suffers from higher power consumption and a more complicated circuit topology when compared with the topologies shown in Figures 9.7 and 9.8.

9.3.2 Current-Mode Instrumentation Amplifier Based on OFCC

Figure 9.10 shows a CMIA based on OFCC. It consists of two operational floating current conveyors (OFCC), two feedback resistors (R_{W1} and R_{W2}), a gain-determined resistor (R_G) and a ground load (R_L), as shown in Figure 9.10. The passive components used in OFCC design are given in Table 9.1. It may be observed that both R_{Z+} and R_{Z-} are very large and therefore may be neglected especially when these resistors act in parallel with a significantly smaller load

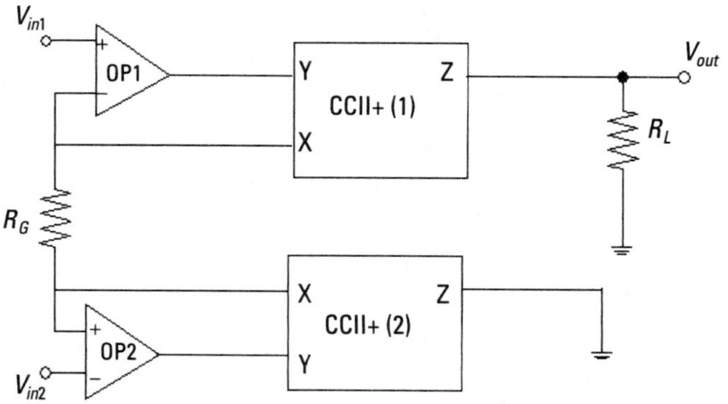

Figure 9.9 Two CCII+ used in conjunction with two op-amps CMIA [3].

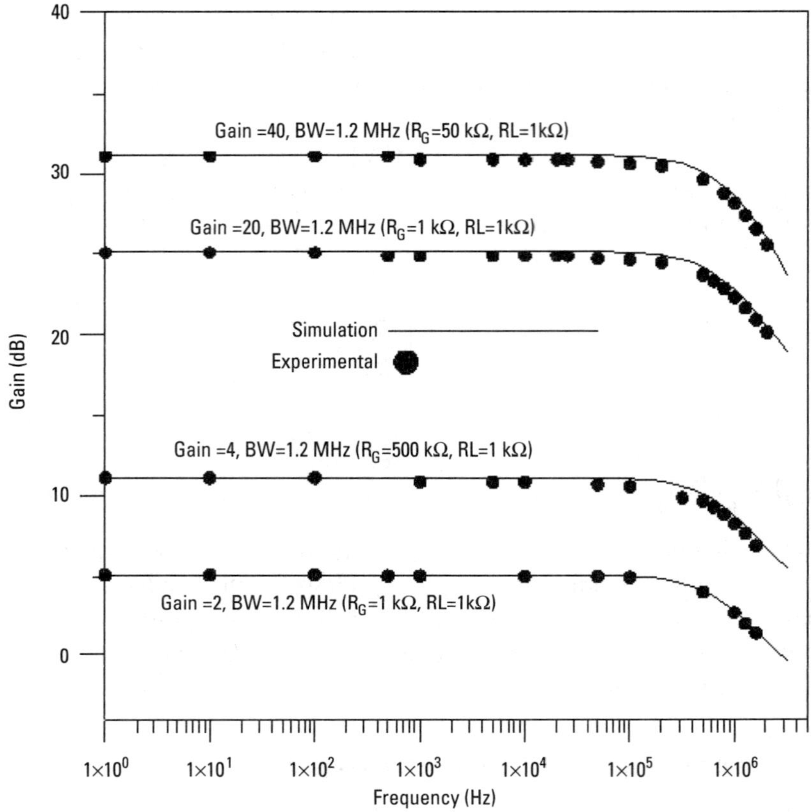

Figure 9.10 The frequency response of the proposed CMIA. (*From:* [9]. Reprinted with permission.)

Table 9.1
The Gain for Different Values of R_G

R_G	Gain
50Ω	40
100Ω	20
500Ω	4
1 kΩ	2

(R_L). Furthermore, the impedance at port Y to ground ($R_Y // C_Y$), as well as port W ($R_T // C_T$) to ground are very high and therefore may be ignored in the frequency range of interest. Similarly, the reactance due to C_X may also be ignored. Lastly, the input resistance at X terminal (R_X) is very small (<1Ω) and hence can be neglected. Figure 9.11 shows the block diagram of the proposed CMIA after introducing the above simplifications.

Taking into consideration both the voltage and current tracking errors of the OFCC, the current tracking error between ports *X*, *Z*+, and *Z*– is:

$$\alpha = 1 - \varepsilon_{+} \tag{9.12}$$

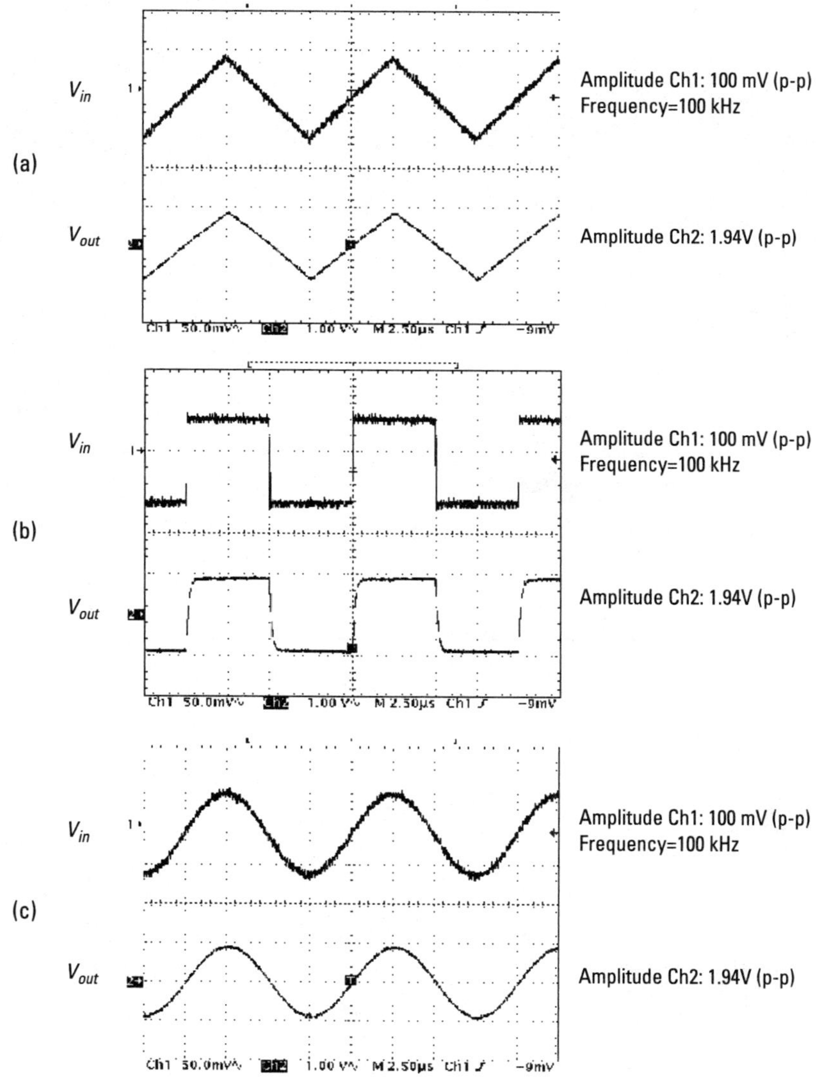

Figure 9.11 The oscilloscope showing the proposed CMIA performance at 100-kHz frequency for $R_G = 100\Omega$ and $R_L = 1\ \text{k}\Omega$ (i.e., gain = 20). (a) Triangle, (b) square, and (c) sinusoidal waveform. (*From:* [9]. Reprinted with permission.)

and

$$\gamma = 1 - \varepsilon_{-} \tag{9.13}$$

where ε_{+} and ε_{-} denote the finite current tracking errors at the high impedance output *Z*+ and *Z*–, respectively. Thus, port currents may then be expressed as $i_{Z+} = \alpha i_X$ and $i_{Z-} = \gamma i_X$.

The voltage tracking error between ports *X* and *Y* is defined as:

$$\beta = 1 - \varepsilon_{v} \tag{9.14}$$

where ε_V denotes the finite voltage tracking error at the low impedance *X* from the high input impedance node *Y*.

The voltage at nodes V_A and V_B, as shown in Figure 9.12, can be expressed as:

$$v_A = \beta_1 v_{in1} \tag{9.15}$$

$$v_B = \beta_2 v_{in2} \tag{9.16}$$

where β_1 and β_2 are the voltage tracking errors of OFCC (1) and OFCC (2), respectively.

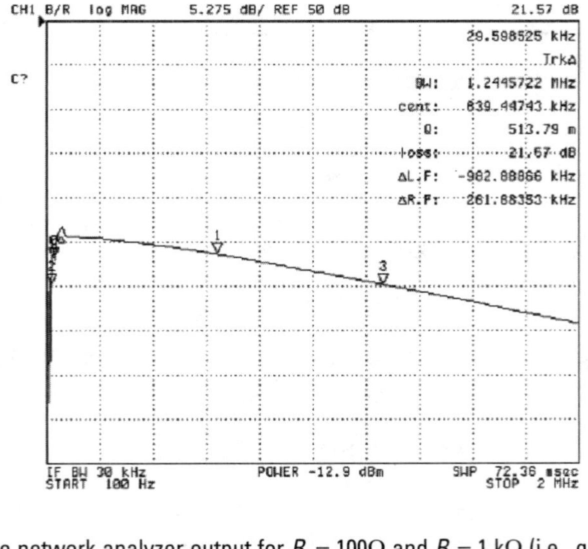

Figure 9.12 The network analyzer output for $R_G = 100\Omega$ and $R_L = 1\ \text{k}\Omega$ (i.e., gain = 20) showing that the $B_W = 1.24$ MHz. (*From:* [9]. Reprinted with permission.)

The resulting current i_x can be calculated as:

$$i_x = \frac{v_A - v_B}{R_G} \tag{9.17}$$

$$i_x = \frac{(\beta_1 v_{in1} - \beta_2 v_{in2})}{R_G} \tag{9.18}$$

Thus, the output terminal currents i_1 and i_2 are calculated as follows:

$$i_1 = \alpha_1 i_x = \frac{\alpha_1 (\beta_1 v_{in1} - \beta_2 v_{in2})}{R_G} \tag{9.19}$$

$$i_2 = \gamma_2 i_x = \frac{\gamma_2 (\beta_1 v_{in1} - \beta_2 v_{in2})}{R_G} \tag{9.20}$$

The output voltage v_o:

$$v_o(s) = (i_1 + i_2)\left(R_L // \frac{1}{sC_Z}\right) \tag{9.21}$$

where $C_Z = C_{Z+} + C_{Z-}$ is the output node capacitance, which is in parallel with R_L.
Using i_1 and i_2 from (9.19) and (9.20), respectively, v_o can now be computed.

$$v_o(s) = \frac{(\beta_1 v_{in1} - \beta_2 v_{in2})(\alpha_1 + \gamma_2) R_L}{R_G (1 + sC_Z R_L)} \tag{9.22}$$

Using (9.22), when $v_{in1} = v_{in2} = v_{cm}$, the common-mode gain ($A_{cm}$) can be written as

$$A_{cm}(S) = \frac{v_o}{v_{cm}} = \frac{(\beta_1 - \beta_2)(\alpha_1 + \gamma_2)R_L}{R_G(1 + sC_Z R_L)} \quad (9.23)$$

For ideal OFCCs, $\alpha_1 = \alpha_2 = \gamma_1 = \gamma_2 = \beta_1 = \beta_2 = 1$; thus, the output voltage is:

$$v_o(s) = \frac{2R_L(v_{in1} - v_{in2})}{R_G(1 + sC_Z R_L)} \quad (9.24)$$

Therefore, the output differential gain A_d can be written as:

$$A_d(s) = \frac{v_o}{v_{in1} - v_{in2}} = \frac{2R_L}{R_G(1 + sC_Z R_L)} \quad (9.25)$$

From (9.25), we can notice that the gain is set by $2R_L/R_G$, and the bandwidth is controlled by $C_Z R_L$. It is also apparent that the differential gain A_d can be adjusted and controlled by R_G without affecting the bandwidth of the CMIA.

9.4 Experimental and Simulation Results of the Proposed CMIA

To verify the operational characteristics of the proposed CMIA, the circuit of Figure 9.11 was simulated using PSPICE version 7.1. The proposed CMIA was also prototyped and the simulation results verified. Each OFCC was constructed using an Analog Devices AD846AQ current feedback op-amp [18] and current mirrors composed of Harris transistor array CA3096CE [19]. The AD846AQ has a bandwidth of 80 MHz at unity gain, and a slew rate of 450 V/μs.

9.4.1 The Differential Gain Measurements

To measure the differential gain of the proposed CMIA, the input voltage was connected to v_{in1} and v_{in2} was connected to the ground. Resistors R_{W1}, R_{W2}, and R_L were set at 1 kΩ and R_G was tested at different values. Table 9.2 shows the values of R_G and the corresponding gain based on (9.25). All resistors have 1% tolerance. Figure 9.13 shows both the experimental as well as the simulation results of the differential gain (A_d) against frequency for R_L kept constant at 1 kΩ and R_G varied to achieve the gain variation. From Figure 9.13, it can be observed that the experimental results validate the simulated results and the analytical results of (9.25), except at frequencies approaching the bandwidth of the OFCC.

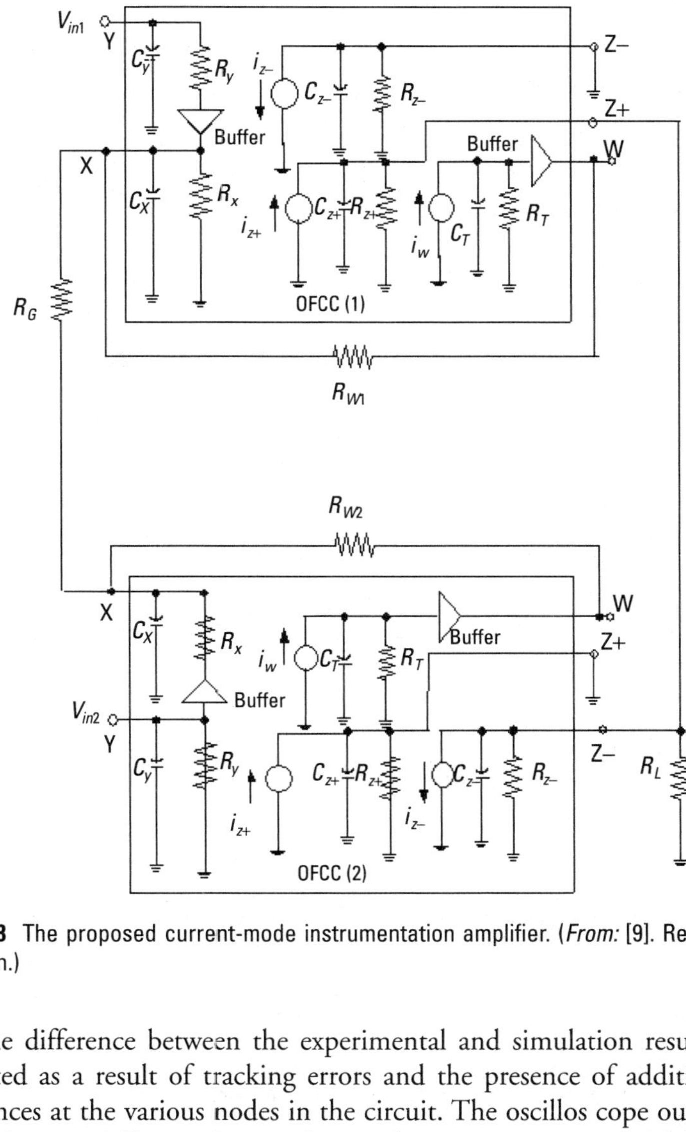

Figure 9.13 The proposed current-mode instrumentation amplifier. (*From:* [9]. Reprinted with permission.)

The difference between the experimental and simulation results can be interpreted as a result of tracking errors and the presence of additional stray capacitances at the various nodes in the circuit. The oscillos cope output taken at 100 kHz at a differential gain of 20 is shown in Figure 9.14. This figure illustrates the behavior of the proposed CMIA when excited by triangular, square, and sinusoidal sources. Figure 9.15 shows the output of the network analyzer, which was used to verify the bandwidth of the proposed circuit at a gain of 20. This figure shows that the measured small signal bandwidth is 1.24 MHz.

9.4.2 Common-Mode Rejection Ratio Measurements

To measure the CMRR of the circuit in Figure 9.11, both v_{in1} and v_{in2} are connected together to the same input voltage source. CMRR was measured experi-

Table 9.2
OFCC Parameters (AD846AQ and CA3096CE)

CFB AD846 Parameters	Current-Mirror Parameters
$R_X = 50\Omega$, $R_Y = 50$ kW	$R_{Z+} = R_{Z-} = 5\ \text{M}\Omega$
CX = 2 pF, CY = 2 pF	CZ+ = CZ– = 6 pF
$R_T = 200\ \text{M}\Omega$, $C_T = 7$ pF	—
$v_n = 2\ \text{nV}/\sqrt{Hz}\ \vert$	—
$i_{mn} = 20\ \text{pA}/\sqrt{Hz}\ \vert$	—
$i_{np} = 6\ \text{pA}/\sqrt{Hz}\ \vert$	—

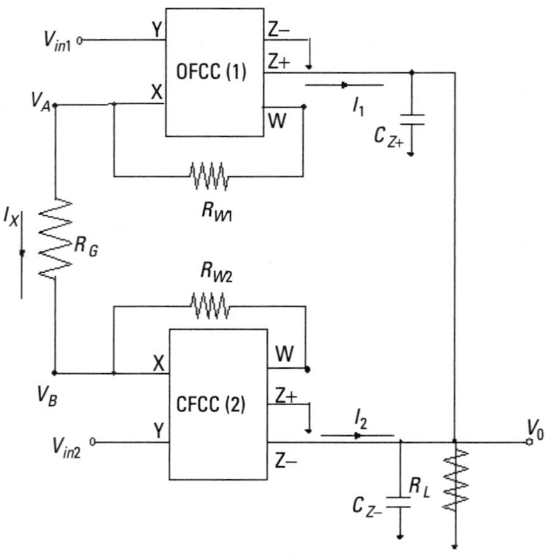

Figure 9.14 The proposed CMIA circuit after simplification. (*From:* [9]. Reprinted with permission.)

mentally as a function of frequency for differential voltage gains of 2, 4, 20, and 40, respectively. The result obtained is plotted in Figure 9.15. From this figure, it can be seen that the proposed topology exhibits a CMRR magnitude of 76 dB and bandwidth of 185 kHz, which are independent of gain. This is larger than the CMRR for circuits proposed in [3, 26, 29] and shown in Figure 9.16.

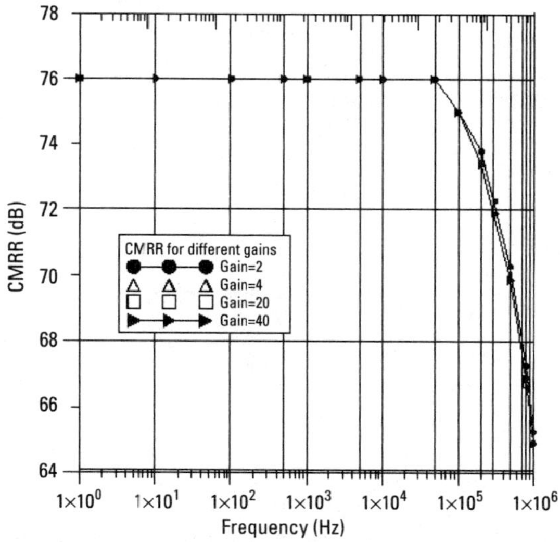

Figure 9.15 CMRR for different gain values. (*From:* [9]. Reprinted with permission.)

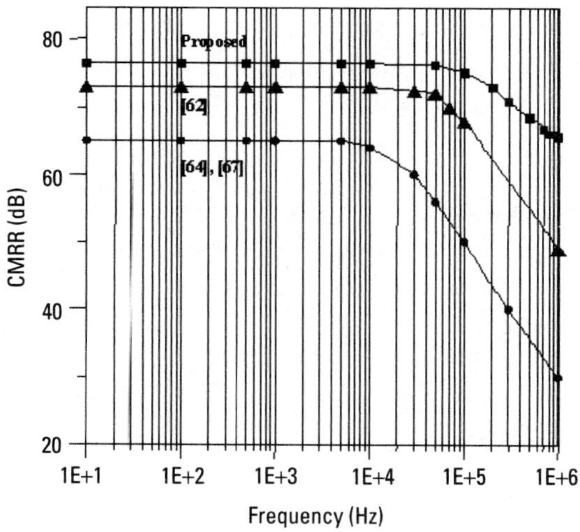

Figure 9.16 CMRR for different CMIA circuits (differential gain = 20). (*From:* [9]. Reprinted with permission.)

9.4.3 Other Features of the Proposed CMIA

Other dynamic and static characteristics of the proposed CMIA have been measured. The proposed CMIA achieves a settling time of 180 ns to 0.01% for a step input for gains of 2 to 40, with a 395 V/μs slew rate. The input offset voltage is 90 μV with gain = 40. A summary of these results is shown in Table 9.3. Us-

Table 9.3
The Dynamic and Static Characteristics of the Proposed CMIA

Characteristics	Value	Condition
Settling time	180 ns	To 0.01% for a step input for gains of 2 to 40
Input offset voltage	90 μV	Gain = 40
Slew rate	395 V/μs	
Bandwidth	1.2 MHz	Independent of gain
CMRR	76 dB	With –3-dB frequency = 185 kHz, it is independent of gain

Source: [9].

ing this CMIA, a high slew rate, high settling time, and low input offset voltage can be obtained. The current feedback operational amplifier (AD846AQ) is the main reason for these features, as it has a fast settling time, high slew rate, and low offset voltage, 110 ns, 450 V/μs, and 75 μV, respectively [18]. Also, it can be observed that the settling time and slew rate of the proposed CMIA are low when compared with the AD846AQ. The reason is that a set of current mirrors has been connected to achieve the OFCC, thus the parasitic capacitance is increased and consequently the settling time and slew rate are decreased.

9.4.4 Noise Results

Figure 9.17 shows a simple model for representing the internal noise effects of the OFCC. For the AD846AQ the noise phenomena are represented in terms of equivalent input noise voltage and current sources, namely, $\overline{v_n^2}$, $\overline{i_{nm}^2}$ and $\overline{i_{np}^2}$.

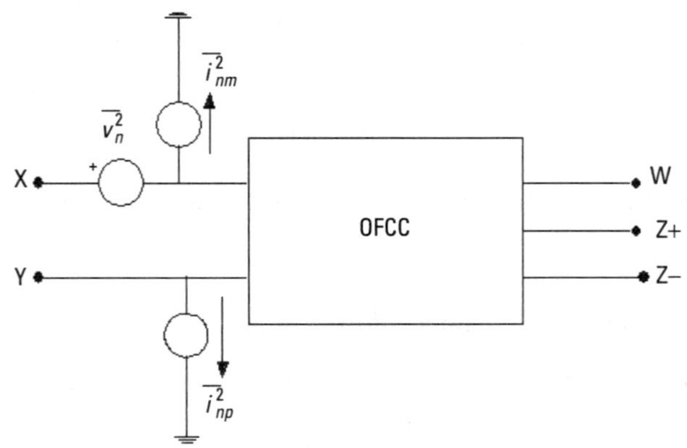

Figure 9.17 Simplified noise model of the OFCC.

The typical values of these noise sources are listed in Table 9.4 [16]. In order to have an insight of the noise performance of the proposed CMIA, the circuit in Figure 9.10 is redrawn in Figure 9.18 with the OFCC noise model shown in Figure 9.17. This circuit has been studied using PSPICE simulation to obtain the noise performance. The values of the noise sources are listed in Table 9.1. The simulation results are plotted in Figure 9.19 for an equivalent resistance of the external input voltage source equal to 1 kΩ (i.e., $R_S = 1$ kΩ, $R_L = R_{W1} = R_{W2} = 1$ kΩ, and $R_G = 50$, 100, 500, and 1 kΩ, respectively). From this figure, it can be observed that the noise level is low; thus, the signal-to-noise ratio, and consequently the dynamic range, will be high.

9.5 Comparison Between Different CMIAs

Table 9.4 shows a performance comparison between the proposed and published CMIAs [3, 26, 29]. It can be seen that the two CCII+ CMIA [29] have a low-frequency differential gain ≈ $R_L/R_G + 2R_X$. Conversely, the enhancement CMIA, which uses two CCII+ and two op-amps [3] has a more accurate low frequency differential gain $A_d \approx R_L/R_G$, which is independent of R_X. However, its bandwidth depends on R_X as shown in (9.9). Also, this topology uses 2 CCII+ in conjunction with 2 op-amps, which results in more power consumption. Both the CCII+ and the CCII+/2 op-amp CMIAs provide a wider bandwidth, which is dependent on the gain. The three CCII+ CMIA [26] provides a low frequency

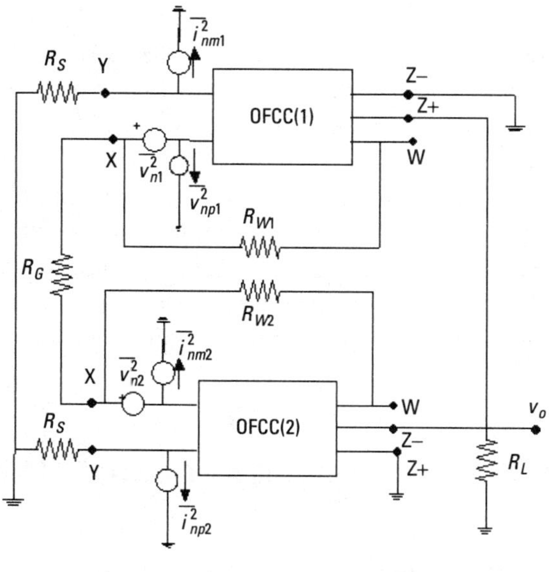

Figure 9.18 Equivalent circuit for analyzing OFCC noise effects on proposed CMIA.

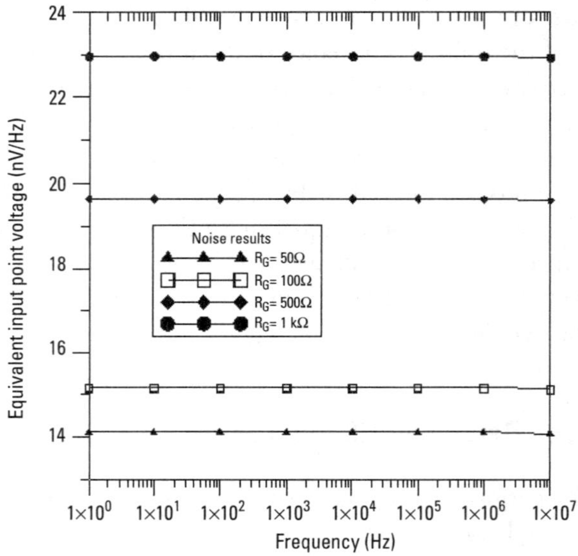

Figure 9.19 Input noise spectral density versus frequency for different R_G. (*From:* [9]. Reprinted with permission.)

differential gain of $2R_L/(R_G + 2R_X)$, which is twice the gain reported in [3, 26] CMIAs. Also, the bandwidth is independent of the gain. However, the proposed topology uses only two OFCCs. It has a low-frequency differential gain $A_d = 2R_L/R_G$, which is also twice that reported in [3, 26]; however, it is similar to the CMIA published in [29]. Moreover, it is independent of R_X and the gain is independent of the bandwidth.

Also from Table 9.4, it can be observed that the proposed CMIA circuit has a higher CMRR, as well a higher bandwidth associated with this CMRR than all the other topologies, and it has a higher gain-bandwidth product with respect to [26, 29].

Finally, the noise performance of the CMIAs proposed in [3, 26, 29] using the CCII's noise model [26] has been studied under the PSPICE environment, and the results are shown in Table 9.3. From Table 9.3, it can be noticed that the proposed CMIA has a lower input noise voltage than the other topologies in [3, 26, 29]. Thus, it has a wider dynamic range and higher signal to noise ratio when compared with [3, 26, 29].

9.6 Testing the Readout Circuit with the Electric Field Based Lab-on-a-Chip

The proposed CMIA was used as a readout circuit for the electric field based lab-on-a-chip (refer to Chapter 8). Figure 9.20 shows a schematic representa-

Table 9.4
Comparison Between the Proposed and Other CMIA Circuits

	Differential Gain for RL/RG = 10				CMRR for RL/RG = 10			
CMIA Reference	**Magnitude (Value)**	**–3-dB Frequency (Bandwidth)**	**Gain Bandwidth Product (MHz)**	**Gain Varies with BW**	**Magnitude (dB)**	**–3-dB Fequency Bandwidth**	**Input Noise Voltage $nV/\sqrt{Hz}$ for RL/RG = 10**	**Number of Building Blocks Used**
[29]	9.09	2 MHz	18.18	Yes	65	16 kHz	20.6	2 CCII
[3]	10	2 MHz	20	Yes	65	16 kHz	22.4	2 CCII 2 op-amp
[26]	17.8	1.4 MHz	24.92	No	73	65 kHz	19.8	3 CCII
Proposed circuit	20	1.2 MHz	24	No	76	185 kHz	15.2	2 OFCC

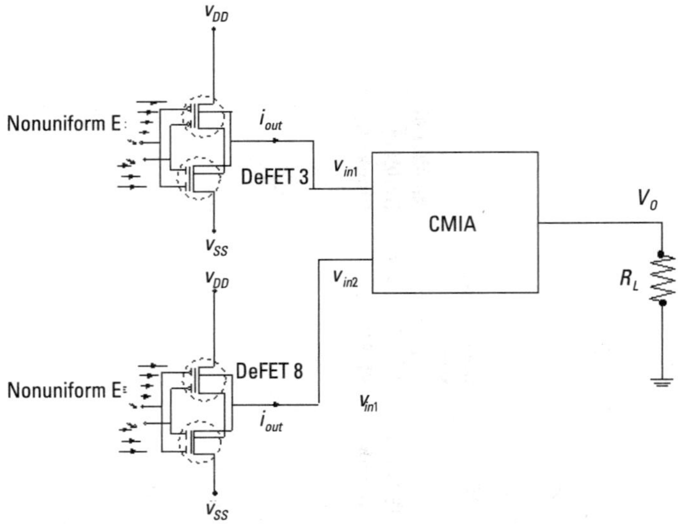

Figure 9.20 Schematic representation of the imager in conjunction with the readout circuit.

tion of the proposed sensors in conjunction with the CMIA. The electrodes of the imager were connected as follows: electrode 2 and electrode 4 are connected to +5V peak-to-peak sinusoidal signal, and electrode 1 and electrode 3 to −5V peak-to-peak; the frequency is 1 MHz. The measured values at the output of DeFET 3 and 8 are plotted in Figure 9.21; these sensors have very close output voltages that cannot be distinguished with direct measurement. Then the outputs of DeFET 3 and 8 were connected to the two inputs (v_{in1} and v_{in2}) of the CMIA, respectively, and the output voltage of the CMIA was measured. The gain of the CMIA is 40. The result is shown in Figure 9.21.

From Figure 9.21, it can be observed that there is about 0.08 mV between DeFET 3 and 8; this value is difficult to be measured directly. However, by using the readout circuit, the differential signal is amplified, the common mode signal is rejected, and the accuracy of the imager is improved. So, the imager can be used in conjunction with the proposed CMIA in biomedical applications that needs high accuracy [21]. A picture of the printed circuit board (PCB) containing both the imager and the CMIA is shown in Figure 9.22.

The integration of the proposed CMIA using CMOS technology on a single chip will be a direction for future work, including the integration of the imager on a single chip to produce a fully integrated lab-on-a-chip.

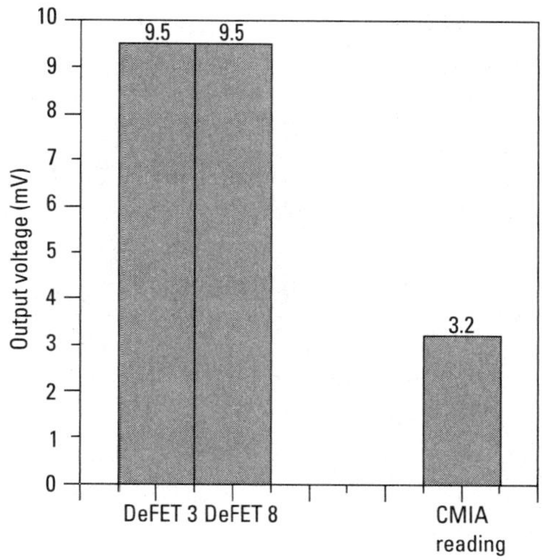

Figure 9.21 The readings of the imager sensors with and without the readout circuit.

Figure 9.22 The PCB contains both the imager and the CMIA.

References

[1] Toumazou, C., F. Lidgey, and D. G. Haigh, "Analog IC Design: The Current-Mode Approach," *IEE Circuits and Systems,* Series 2, 1993.

[2] Tomazou, C., F. J. Lidgey, and P. Y. K. Cheng, "Current-Mode Analogue Signal Processing Circuits—A Review of Recent Developments," *IEEE International Symposium on Circuits and Systems,* Vol. 3, 1989, pp. 1572–1757.

[3] Gift, S. J. G., "An Enhanced Current-Mode Instrumentation Amplifier," *IEEE Trans. on Instrumentation and Measurement,* Vol. 50, No. 1, 2001, pp. 85–88.

[4] Gkotsis, I., G. Souliotis, and I. Haritantis, "Instrumentation Amplifier Based Analogue Interface," *International Conference on Electronic Circuits and Systems (ICECS 98),* Vol. 1, Lisbon, Portugal, 1998, pp. 317–320.

[5] Koli, K., and K. A. I. Halonen, "CMRR Enhancement Techniques for Current-Mode Instrumentation Amplifiers," *IEEE Trans. on Circuits and Systems*, Vol. 47, No. 5, 2000, pp. 622–632.

[6] Zhu, Q. S., F. J. Lidgey, and W. J. Su, "High CMRR, Second Generation Current-Mode Instrumentation Amplifier," *International Symposium on Circuits and Systems (ISCAS 93)*, Chicago, IL, Vol. 2, 1993, pp. 1326–1328.

[7] Wilson, B., "Recent Developments in Current Conveyors and Current-Mode Circuits," *IEE Proceedings on Electronic Circuits and Systems*, Vol. 137, No. 12, 1990, pp. 63–77.

[8] Sedra, A. S., G. W. Roberts, and F. Gohn, "The Current Conveyor: History, Progress and New Results," *IEE Proceedings on Electronic Circuits and Systems*, Vol. 137, No. 12, 1990, pp. 78–87.

[9] Ghallab, Y. H., et al., "A Novel Current-Mode Instrumentation Amplifier Based on Operational Floating Current Conveyor," *IEEE Trans. on Instrumentation and Measurement*, Vol. 54, No. 5, October 2005, pp. 1941–1994.

[10] Khan, A., M. Al-Turiaia, and M. Abo El-Ela, "Operational Floating Current Conveyor: Characteristics, Modeling and Applications," *IEEE Instrumentation and Measurement Technical Conference (IMTC94)*, Hamamatsu, Japan, 1994, pp. 788–790.

[11] Ghallab, Y. H., M. Abo El-Ela, and M. Elsaid, "Operational Floating Current Conveyor: Characteristics, Modeling and Experimental Results," *International Conference on Microelectronics (ICM99)*, Kuwait, 1999, pp. 59–62.

[12] Ghallab, Y. H., and W. Badawy, "A Novel pH Sensor Current Mode Read-Out Circuit Using Only Two Operational Floating Current Conveyor," Biomedical *Circuits and Systems (BioCAS 2004)*, Singapore, 2004, pp. S1.5-13–S1.5-16.

[13] Ghallab, Y. H., and W. Badawy, "A Novel pH Sensor Current Mode Read-Out Circuit Using Operational Floating Current Conveyor," *Proceedings of International Conference on MEMS, NANO and Smart Systems (ICMENS 2004)*, Banff, Canada, 2004, pp. 262–265.

[14] Ghallab, Y. H., W. Badawy, and K. V. I. S. Kaler, "A Novel Differential ISFET Current Mode Read-Out Circuit Using Operational Floating Current Conveyor," *Proceedings of International Conference on MEMS, NANO and Smart Systems (ICMENS 2003)*, Banff, AB, Canada, 2003, pp. 255–258.

[15] Ghallab, Y. H., et al., "A New Second-Order Active Universal Filter with Single Input and Three Outputs Using Operational Floating Current Conveyor," *International Conference on Microelectronics (ICM 02)*, Beirut, Lebanon, 2002, pp. 42–45.

[16] Ghallab, Y. H., M. Abo El-Ela, and M. Elsaid, "A Novel Universal Voltage-Mode Filter with Three Inputs and Single Output Using Only Two Operational Floating Current Conveyor," *Proceedings of the International Conference on Microelectronics (ICM 00)*, Tehran, Iran, 2000, pp. 95–98.

[17] Soclof, S., Chapter 9 in *Design and Applications of Analog Integrated Circuits*, Englewood Cliffs, NJ: Prentice-Hall, 1991, pp. 443–460.

[18] Analog Devices, *450 V/μs, Precision, Current-Feedback Opamp (AD846)*, pp. (2-307)–(2-317).

[19] Harris Semiconductor, *CA3096, CA3096A, CA3096C, NPN Transistor Arrays*, File Number 595.4, December 1997.

[20] Azhari, S. J., and H. Fazalipoor, "A Novel Current-Mode Instrumentation Amplifier (CMIA) Topology," *IEEE Trans. on Instrumentation and Measurement*, Vol. 49, No. 6, 2000, pp. 1272–1277.

[21] Webster, J. G., *Medical Instrumentation: Application and Design*, 3rd ed., New York: John Wiley & Sons, 1998.

[22] Webster, J. G., *Bioinstrumentation*, New York: John Wiley & Sons, 2004.

[23] Szynowski, J., "CMRR Analysis in Instrumentation Amplifiers," *Electronics Letters*, Vol. 19, No. 14, 1983, pp. 547–549.

[24] Areny, R. P., and J. G. Webster, "Common-Mode Rejection Ratio in Differential Amplifier Stages," *IEEE Trans. on Instrumentation and Measurement*, Vol. 40, No. 4, 1991, pp. 669–676.

[25] Areny, R. P., and J. G. Webster, "Common Mode Rejection Ratio for Cascaded Differential Amplifier Stages," *IEEE Trans. on Instrumentation and Measurement*, Vol. 40, No. 4, 1991, pp. 677–681.

[26] Khan, A. A., M. A. Al-Turaigi, and M. Abou El-Ela, "An Improved Current-Mode Instrumentation Amplifier with Bandwidth Independent of Gain," *IEEE Trans. on Instrumentation and Measurement*, Vol. 44, No. 4, 1995.

[27] Galanis, C., and I. Haritantis, "An Improved Current-Mode Instrumentation Amplifier," *International Conference on Electronic Circuits and Systems (ICECS 96)*, Rodos, Greece, Vol. 1, 1996, pp. 65–68.

[28] Kaulberg, T., "A CMOS Current-Mode Operational Amplifier," *IEEE Journal of Solid State Circuits*, Vol. 28, No. 7, 1993, pp. 849–852.

[29] Wilson, B., "Universal Conveyor Instrumentation Amplifier," *Electronics Letters*, Vol. 25, No. 7, 1989, pp. 470–471.

[30] Toumazou, C., and F. J. Lidgey, "Novel Current-Mode Instrumentation Amplifier," *Electronic Letters*, Vol. 25, No. 3, 1989, pp. 228–230.

[31] Azhari, S. J., and H. Fazalipoor, "A Novel Current-Mode Instrumentation Amplifier (CMIA) Topology," *IEEE Trans. on Instrumentation and Measurement*, Vol. 49, No. 6, 2000, pp. 1272–1277.

10

Current-Mode Wheatstone Bridge for Lab-on-a-Chip Applications

10.1 Introduction

The voltage-mode Wheatstone bridge (VMWB) offers a good method for measuring small resistance changes accurately. Therefore, it is used for sensing temperature, strain, pressure, fluid flow, and dew point humidity [1]. It consists of four resistors [2] (shown in Figure 10.1), and the output voltage can be computed as:

$$V_o = \left(\frac{R_2}{R_1 + R_2} - \frac{R_4}{R_3 + R_4} \right) V_{in} \tag{10.1}$$

The null condition exists when V_o is zero. The null condition occurs when

$$R_1 R_4 = R_2 R_3 \tag{10.2}$$

Assume that

$$R_1 = R_4 = R_o \mp \Delta R \tag{10.3}$$

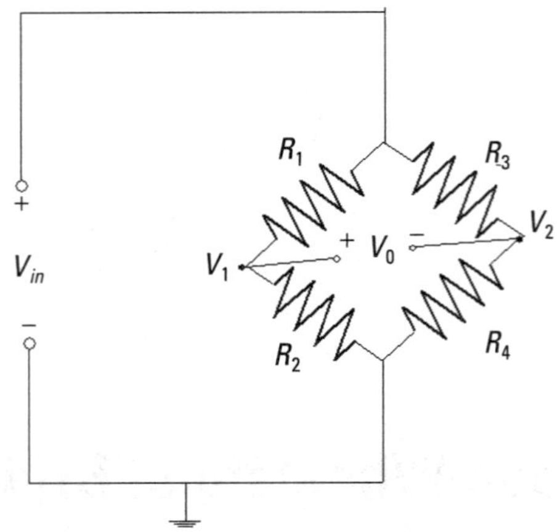

Figure 10.1 Traditional voltage-mode Wheatstone bridge.

$$R_2 = R_3 = R_o \pm \Delta R \tag{10.4}$$

Thus, V_o is proportional to ΔR as follows:

$$V_o = V_1 - V_2 = \pm \frac{\Delta R}{R_o} \cdot V_{in} \tag{10.5}$$

and

$$V_{in} = V_1 + V_2 \tag{10.6}$$

From a current-mode point of view, a similar formula as (10.5) can be extracted [3]:

$$I_{out} = \pm \frac{\Delta G}{G_o} \cdot I_{in} \tag{10.7}$$

where $G_o = 1/R_o$ and $\Delta G = 1/\Delta R$.

A method based on circuit duality concept [4–11] has been developed to develop a current-mode Wheatstone bridge (CMWB) [3]. The advantages of the OFCC-based CMWB are: (1) the reduction of sensing passive elements (i.e., two resistors can be used instead of four to achieve the same performance), (2) the ability of the superposition principle to be applied without adding any signal conditioning circuitry, thus making possible the addition of sensor effects, and (3) a higher common-mode cancellation.

In [3], two different topologies to implement a CMWB have been proposed. The first one uses two second-generation current conveyors (CCII), a positive CCII+ and a negative CCII– (see Figure 10.2). Figure 10.3 shows the equivalent circuit of Figure 10.2, taking into consideration the equivalent input resistance at the *X* terminal (R_x) of the CCIIs. Assuming that $R_1 = R_o \mp \Delta R$ and $R_2 = R_o \pm \Delta R$, and by using a routine circuit analysis, we can prove that the output current I_o is related to the reference current I_{ref} by (10.8):

$$I_{out} = I_1 - I_2 = \frac{\pm \Delta R}{R_o + R_x} \cdot I_{ref} \tag{10.8}$$

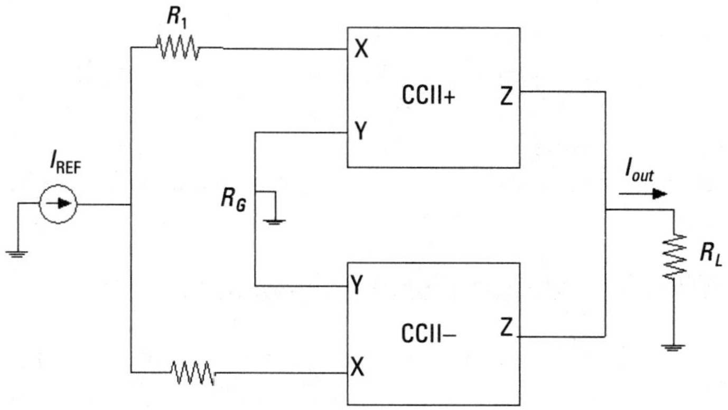

Figure 10.2 The CMWB based on CCII. (Reprinted from [3] with permission.)

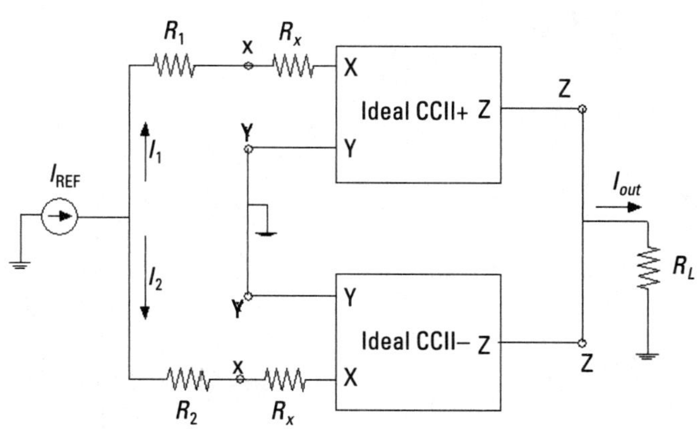

Figure 10.3 Practical CMWB based on the equivalent circuit of CCII.

From (10.8), it can be observed that I_{out} is inversely proportional to R_x, and the accuracy is limited by the tolerance of R_x, which is low. Also, the linearization topology used, when only one resistor is varying (i.e., $R_1 = R_o \mp \Delta R$ and $R_2 = R_o$), is shown in Figure 10.4. In this case the output current I_o is related to the reference current I_{ref} by:

$$I_o = I_x = \frac{\pm \Delta R + R_x}{R_o + R_x} \cdot I_{ref} \tag{10.9}$$

Therefore, the disadvantages of using CCII as a building block in a CMWB are the limited accuracy [see (10.8) and (10.9)] and the need of more circuitry for linearization.

The second approach used in [3] to implement a CMWB using operational floating current conveyor (OFCC) is shown in Figure 10.5; this topology has the same advantages mentioned earlier. Moreover, it has a higher accuracy, as I_o does not depend on R_x. However, there is no reduction of the sensing resistors, since it uses two excess resistors (i.e., two R_3 resistors). Furthermore, there is neither experimental nor simulation results to verify the topologies proposed in [3].

10.2 CMWB Based on Operational Floating Current Conveyor

The operational floating current conveyor (OFCC)-based CMWB consists of three OFCCs, three feedback resistors (R_{W1}, R_{W2}, and R_{W3}), two sensitive resistors (R_1 and R_2) (one or both of them can represent a sensor), and a ground load (R_L),

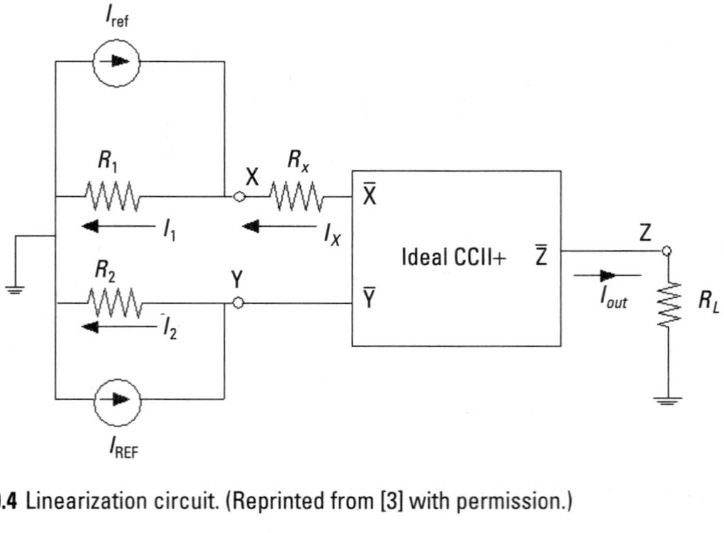

Figure 10.4 Linearization circuit. (Reprinted from [3] with permission.)

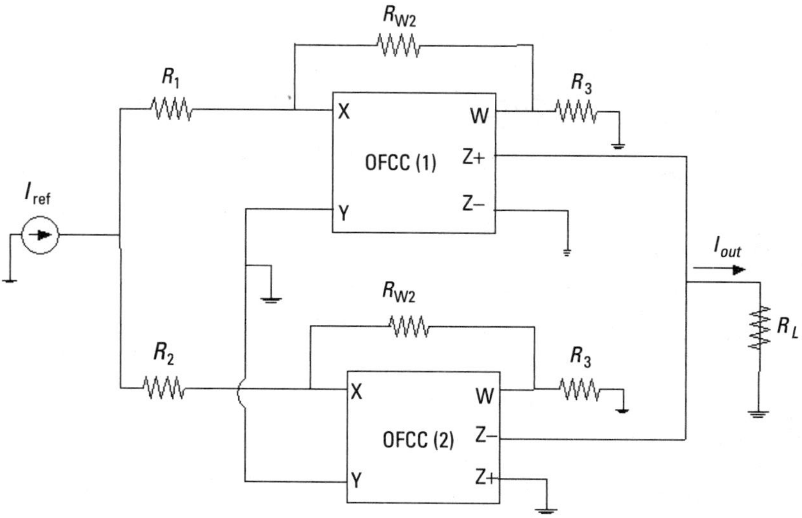

Figure 10.5 The CMWB-based OFCC. (Reprinted from [3] with permission.)

as shown in Figure 10.6. The passive components used in the OFCC design are given in Table 10.1. It may be observed that both R_{Z+} and R_{Z-} are very large and therefore may be neglected, especially when these resistors act in parallel with a significantly smaller load (R_L). Furthermore, the input resistance at X terminal (R_X) is very small (<0.002Ω) and hence can be neglected.

Taking into consideration both the voltage and current tracking errors of the OFCC, the current tracking error between ports X, $Z+$, and $Z-$ is:

$$\alpha = 1 - \varepsilon_+ \tag{10.10}$$

and

$$\gamma = 1 - \varepsilon_- \tag{10.11}$$

where ε_+ and ε_- denote the finite current tracking errors at the high-impedance output $Z+$ and $Z-$, respectively. Thus, port currents may then be expressed as $I_{z-} = \alpha I_x$ and $I_{z-} = \gamma I_x$. There is a voltage tracking between the X and Y terminals of the OFCC, so $V_A = V_B = 0$.

The currents I_1 and I_2 are calculated as follows:

$$I_1 = \frac{R_2}{R_1 + R_2} \cdot I_{in} \tag{10.12}$$

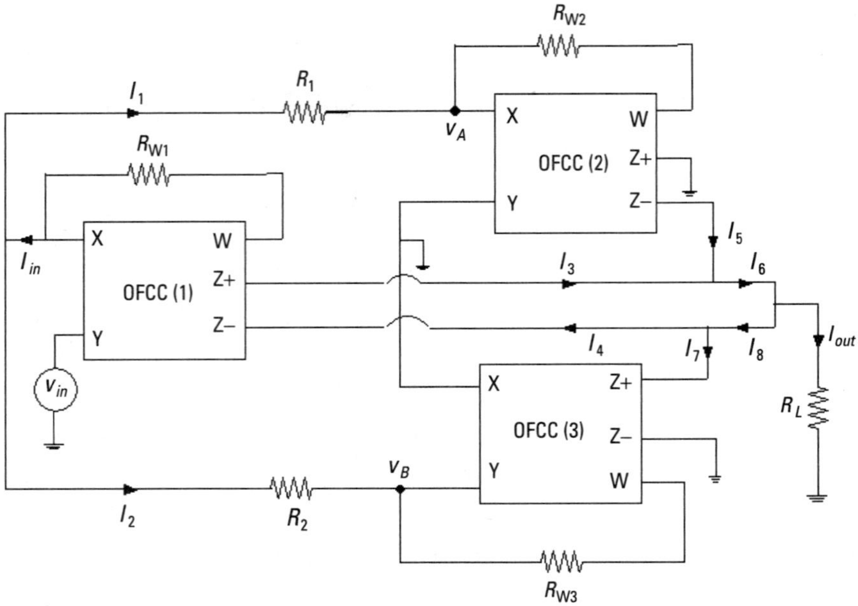

Figure 10.6 The OFCC-based CMWB. (Reprinted from [12] with permission.)

Table 10.1
OFCC DC Parameters (AD846AQ and CA3096CE)

CFB AD846 Parameters	Current-Mirror Parameters
$R_X = 50\Omega$, $R_Y = 50$ kΩ	$R_{Z+} = R_{Z-} = 5$ MΩ
$R_T = 200$ MΩ	

$$I_2 = \frac{R_1}{R_1 + R_2} \cdot I_{in} \tag{10.13}$$

The input current I_{in} is:

$$I_{in} = I_1 + I_2 \tag{10.14}$$

The output terminals currents from the OFCC (1), I_3 and I_4, can be calculated as follows:

$$I_3 = \alpha_1 I_{in} \tag{10.15}$$

$$I_4 = \gamma_1 I_{in} \tag{10.16}$$

The output current from the OFCC (2), I_5, is:

$$I_5 = \gamma_2 I_1 \tag{10.17}$$

From (10.15), and (10.17), I_6 can be obtained

$$I_6 = I_3 + I_5 = \alpha_1 I_{in} + \gamma_2 I_1 \tag{10.18}$$

The output current from the OFCC (3), I_7, is computed:

$$I_7 = \alpha_3 I_2 \tag{10.19}$$

The summation of I_3 and I_7, which is I_8, can be obtained from (10.16) and (10.19):

$$I_8 = I_4 + I_7 + \gamma_1 I_{in} + \alpha_3 I_2 \tag{10.20}$$

The output current (I_{out}) is the difference between I_8 and I_6; therefore, we can calculate I_{out} from (10.18) and (10.20):

$$I_{out} = I_6 - I_8 \tag{10.21}$$

$$I_{out} = \alpha_1 I_{in} + \gamma_2 I_1 - \gamma_1 I_{in} - \alpha_3 I_2 \tag{10.22}$$

For ideal OFCCs, $\alpha_1 = \alpha_2 = \alpha_3 = \gamma_1 = \gamma_2 = \gamma_3 = 1$. Thus, the output current is:

$$I_{out} = I_1 - I_2 \tag{10.23}$$

From (10.7) and (10.8) into (10.23), we can get I_{out} in terms of R_1 and R_2:

$$I_{out} = \frac{R_2 - R_1}{R_1 + R_2} \cdot I_{in} \tag{10.24}$$

Thus, if we have $R_1 = R_o \mp \Delta R$ and $R_2 = R_o \pm \Delta R$, then I_{out}

$$I_{out} = \frac{\pm \Delta R}{R_o} \cdot I_{in} \qquad (10.25)$$

From (10.25), a linear relationship can be observed between I_{out} and ΔR, and it does not depend on R_x [see (10.8)]. Therefore, the accuracy is improved when compared with the CMWB-based CCII in [3]. Also, from (10.23), we can observe that the OFCC-based topology has a high common-mode rejection ratio (CMRR) for the common-mode current (i.e., $I_1 = I_2$). In other words, when R_1 is equal to R_2, then I_1 is equal I_2 (the common-mode current) and the output current $I_{out} = 0$. Also, an output current is obtained; hence, the superposition principle can be applied, as shown in Figure 10.7. Thus, it is possible to add the effects of any number of sensors without adding any other circuitry, and this is the great advantage over the voltage-mode Wheatstone bridge VMWB. Moreover, we are using only two sensitive resistors to get the same performance, instead of four as in traditional VMWB.

10.3 A Linearization Technique Based on an Operational Floating Current Conveyor

In case only one resistor is sensitive to the variation of measuring (i.e., $R_1 = R_o$ and $R_2 = R_o \mp \Delta R$), the resulting output current in Figure 10.6 causes some nonlinearity, which can be expressed using (10.24) as:

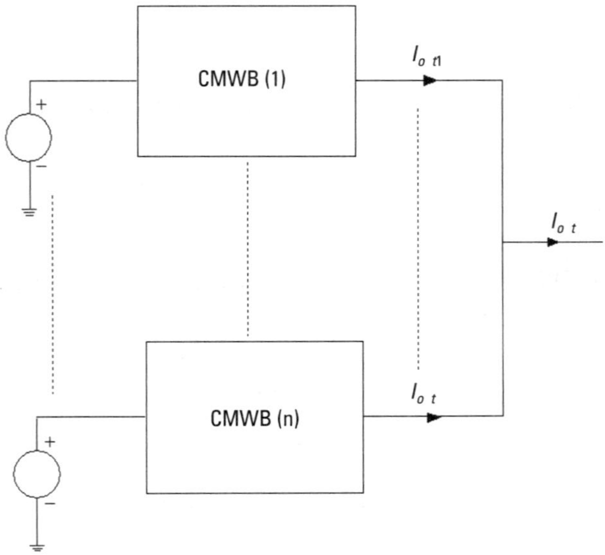

Figure 10.7 The superposition principle can be applied with the OFCC-based CMWB.

$$I_{out} = \pm\frac{\Delta R}{2R_o + \Delta R}\cdot I_{in} \approx \pm\frac{\Delta R}{2R_o} \tag{10.26}$$

From (10.25), we can notice that the resulting output current causes some nonlinearity.

To compensate the nonlinearity, another circuit has been proposed based on OFCC, as shown in Figure 10.8. This circuit has the same number of OFCCs as the OFCC-based CMWB (i.e., three OFCCs; see Figure 10.6). Thus, no additional circuitry is needed to implement it, and a reconfiguration is only needed to compensate the nonlinearity (see Figure 10.8).

Taking into consideration the current tracking error, see (10.10) and (10.11), as well as the voltage tracking error between ports *X* and *Y*:

$$\beta = 1 - \varepsilon_V \tag{10.27}$$

where ε_V denotes the finite voltage tracking error at the low impedance *X* from the high input impedance node *Y*.

The output currents of OFCC (1), I_1 and I_2, are shown in (10.28) and (10.29), respectively:

$$I_1 = \alpha_1 I_{in} \tag{10.28}$$

$$I_2 = \gamma_1 I_{in} \tag{10.29}$$

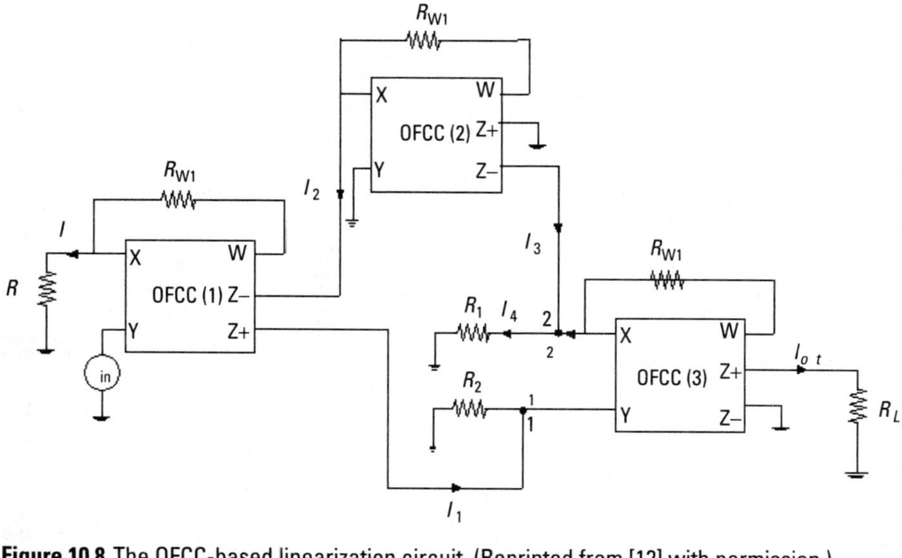

Figure 10.8 The OFCC-based linearization circuit. (Reprinted from [12] with permission.)

The output current of OFCC (2), I_3, is:

$$I_3 = \alpha_2 I_2 \tag{10.30}$$

From (10.29) into (10.30),

$$I_3 = \alpha_2 \gamma_1 I_{in} \tag{10.31}$$

At node 1, the voltage V_1 can be calculated as:

$$V_1 = I_1 R_2 \tag{10.32}$$

From (10.28) into (10.32),

$$V_1 = \alpha_1 I_{in} R_2 \tag{10.33}$$

The voltage at node 2, V_2, is related to V_1 by

$$V_2 = \beta_2 V_1 = \beta_2 \alpha_1 I_{in} R_2 \tag{10.34}$$

At node 2:

$$\begin{aligned} I_4 &= \frac{V_2}{R_1} \\ I_4 &= \frac{\beta_2 \alpha_1 R_2 I_{in}}{R_1} \end{aligned} \tag{10.35}$$

Also,

$$I_x = I_4 - I_3 \tag{10.36}$$

From (10.31) and (10.35) into (10.36), we can get a relationship between I_x and I_{in}:

$$I_x = \left(\frac{\beta_2 \alpha_1 R_2}{R_1} - \alpha_2 \gamma_1 \right) I_{in} \tag{10.37}$$

The output current I_{out} is related to I_x by (10.38):

$$I_{out} = \alpha_3 I_x \tag{10.38}$$

From (10.37) into (10.38), I_{out} is related to I_{in} by:

$$I_{out} = \alpha_3 \left(\frac{\beta_2 \alpha_1 R_2}{R_1} - \alpha_2 \gamma_1 \right) I_{in} \tag{10.39}$$

For ideal OFCCs, $\alpha_1 = \alpha_2 = \alpha_3 = \gamma_1 = \beta_2 = 1$. Thus, the output current I_{out} is:

$$I_{out} = \left(\frac{R_2}{R_1} - 1 \right) I_{in} \tag{10.40}$$

As $R_1 = R_o$ and $R_2 = R_o \mp \Delta R$, so I_{out} can be given as:

$$I_{out} = \frac{\mp \Delta R}{R_o} I_{in} \tag{10.41}$$

From (10.41), we can observe that the output current is linearly related to ΔR. Therefore, the circuit used in Figure 10.8 can be used for linearization.

10.4 Experimental and Simulation Results

To verify the operational characteristics of the OFCC-based CMWB, the circuit of Figure 10.7 was simulated using PSPICE version 7.1. The OFCC-based CMWB was also prototyped and the simulation results verified. Each OFCC was constructed using an Analog Devices AD846AQ current feedback op amp [10] and current mirrors composed of the Harris transistor array CA3096CE [11]. The AD846AQ has a bandwidth of 80 MHz at unity gain and a slew rate of 450 V/μs.

10.4.1 The Differential Measurements

To measure the differential characteristics of the OFCC-based CMWB, the input voltage to V_{in} is connected. Resistors R_{w1}, R_{w2}, R_{w3}, R_L, and R_1 were set at 1 KΩ, and R_2 was tested at different values (1.5 KΩ, 2 KΩ, 3 KΩ, and 4 KΩ). All resistors have 1% tolerance. Figure 10.9 shows both the experimental and the simulation results of the output current (I_{out}) against the input voltage (V_{in}) for R_1 kept constant at 1 KΩ and R_2 varied. From Figure 10.9, it can be observed that the experimental results validate the simulated results and the analytical results of

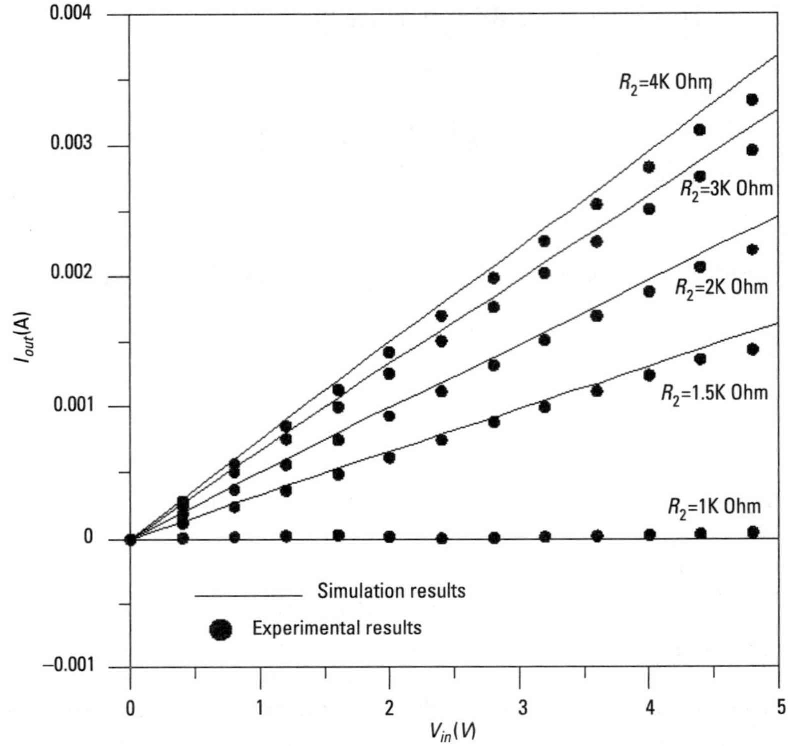

Figure 10.9 The DC response of the OFCC-based CMWB with $R_1 = 1$ KΩ and R_2 varies. (Reprinted from [12] with permission.)

(10.25). The difference between the experimental and simulation results can be interpreted as a result of tracking errors and the tolerance of the resistors.

The AC performance of the OFCC-based CMWB is also tested by connecting an AC source at v_{in} (v_{in} = 200 mV) at different values of R_2. Figure 10.10 shows both the experimental and the simulation results of the i_{out} against frequency for R_1 kept constant at 1 kΩ and R_2 varied. We can observe from this figure that the experimental results are in good agreement with the simulation results, except at frequencies approaching the bandwidth of the OFCC. The difference between the experimental and simulation results can be interpreted as a result of tracking errors and the presence of additional stray capacitances at the various nodes in the circuit. Also, the bandwidth (50 MHz) is high and constant with different R_2 values.

10.4.2 Common-Mode Measurements

To measure the common-mode rejection (CMR) of the circuit in Figure 10.7, R_1 and R_2 are set to 1 KΩ. CMR was measured experimentally as a function of

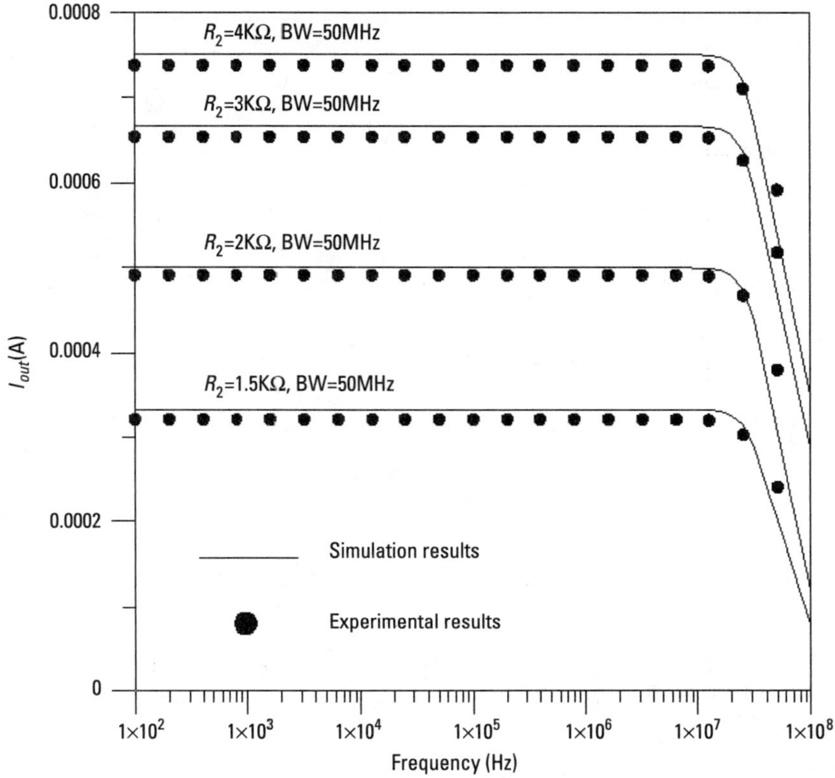

Figure 10.10 The frequency response for the OFCC-based CMWB with R_1 = 1 KΩ and R_2 varied. (Reprinted from [12] with permission.)

frequency. The result obtained is plotted in Figure 10.11. From this figure, the OFCC-based topology exhibits a very small common-mode current, which is independent on the frequency. This current can be interpreted as a result of tracking errors. On the other hand, the common-mode current of the CMWB based on CCII in [3], also shown in Figure 10.11, increases with frequency. This is due to the tracking errors and the changing of the input resistance at the X terminal (R_x) with the frequency.

10.5 Discussion

Table 10.2 shows a comparison between the OFCC-based and the other voltage-mode and current-mode Wheatstone bridges. It can be seen that the advantages of the CMWBs over the conventional VMWBs are as follows. First, the number of sensing passive elements are reduced (i.e., two resistors can be used instead of four), and getting the same performance. Second, the superposition principle can be applied without adding any signal conditioning circuitry. Thus, the addi-

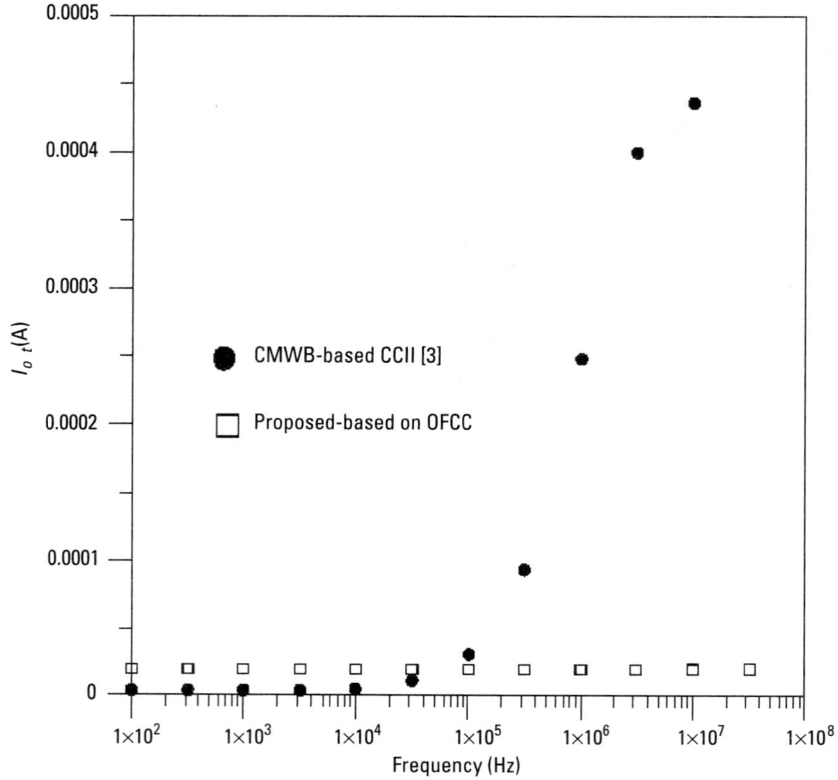

Figure 10.11 Experimental results for $R_1 = R_2 = 1$ KΩ to compare between the CMR of the OFCC-based CMWB and the CMWB based on CCII in [3]. (Reprinted from [12] with permission.)

Table 10.2
Comparison Between the OFCC-Based CMWB and Other VMWB- and CCII-Based CMWB [3]

Comparison Factor	Conventional VMWB	CCII-Based CMWB [3]	OFCC-Based CMWB
1. Number of sensitive elements used	4	2	2
2. Common-mode rejection	High	High	High
3. Suitable to integration	Needs large area	Needs small area	Needs small area
4. Accuracy	High	Low	High
5. Ability to use superposition	Not able	Able	Able
6. Linearization	Needs additional circuitry	Does not need additional circuitry	Does not need additional circuitry

tion of sensor effects is possible, and finally it has a higher common-mode cancellation. On the other hand, the OFCC-based CMWB has a higher accuracy (i.e., it does not depend on R_x) when compared with the CCII-based CMWB, proposed in [3].

References

[1] Perez, R. J., *Design of Medical Electronic Devices,* San Diego, CA: Academic Press, 2002.

[2] Carr, J. J., and J. M. Brown, *Introduction to Biomedical Equipment Technology,* New York: John Wiley & Sons, 1981.

[3] Azhari, S., and H. Kaabi, "AZKA Cell, the Current-Mode Alternative of Wheatstone Bridge," *IEEE Trans. on Circuits and Systems-I,* Vol. 47, No. 9, 2000, pp. 1277–1284.

[4] Huelsman, L. P., *Basic Circuit Theory,* 3rd ed., Upper Saddle River, NJ: Prentice-Hall, 1991.

[5] Khan, A., M. Al-Turiaia, and M. Abo El-Ela, "Operational Floating Current Conveyor: Characteristics, Modeling and Applications," *IMTC94,* Hamamtsu, Japan, 1994, pp. 788–790.

[6] Ghallab, Y. H., et al., "Operational Floating Current Conveyor: Characteristics, Modeling and Experimental Results," *ICM99,* Kuwait, 1999.

[7] Soclof, S., *Design and Applications of Analog Integrated Circuits,* Englewood Cliffs, NJ: Prentice-Hall, Ch. 9, 1991, pp. 443–460.

[8] Azhari, S. J., and H. Fazalipoor, "A Novel Current-Mode Instrumentation Amplifier (CMIA) Topology," *IEEE Trans. on Instrumentation and Measurement,* Vol. 49, No. 6, 2000, pp. 1272–1277.

[9] Gift, S. J. G., "An Enhanced Current-Mode Instrumentation Amplifier," *IEEE Trans. on Instrumentation and Measurement,* Vol. 50, No. 1, 2001, pp. 85–88.

[10] Analog Devices Manual, "450 V/μs, Precision, Current-Feedback OpAmp (AD846)," pp. (2-307)–(2-317).

[11] Harris Semiconductor "CA3096, CA3096A, CA3096C, NPN Transistor Arrays," File Number 595.4, December 1997.

[12] Ghallab, Y. H., and W. Badawy, "A New Topology for a Current-Mode Wheatstone Bridge," *IEEE Trans. on Circuits and Systems II,* Vol. 53, No. 1, January 2006, pp. 18–22.

11

Current-Mode Readout Circuits for the pH Sensor

11.1 Introduction

The application of the ion sensitive field effect transistor (ISFET) device to measure chemical quantities was demonstrated in 1970 [1] and has since been studied extensively [2, 3]. An ISFET is a small-sized device with a rapid response that can be readily fabricated using a conventional MOS process [4]. Unfortunately, there are many challenges facing the ISFET such as the threshold voltage time dependence, the measurement temperature dependence, and the technological difficulties associated with the packing of miniature reference electrodes limit the commercial viability of all ISFET applications intended for the long-term pH monitoring in chemical analyses. In the past, in order to address these challenges, different approaches have been proposed to address sensor design, packaging, and signal compensation. Several solutions include on-chip fabrication of an Ag/AgCl electrode with microfabrication technique, as well as the chemical treatment of the sensing membrane [5, 6].

The alternative approach that uses a differential arrangement, which consists of an ISFET and a reference FET (REFET) with the same ISFET characteristics but insensitive to the pH variation. Both are integrated in the same device structure [7–9].

A differential ISFET sensor connected to a readout circuit can be usefully applied in a variety of applications dealing with chemical measurements as well as biomedical diagnoses. These ISFET measurements network can be voltage [10–13] or current mode of operation. In the voltage operation, the changes in the output voltage values represent the sensing of chemical or biological phenomenon. In the current mode of operation, the changes in the output current values represent the sensed quantities.

The current-mode method is a more promising approach because of its wide dynamic range and bandwidth, lower power consumption, higher usable gain, and its possible integration to the analog-to-digital (A/D) circuitry [14]. There are many devices based the current-mode approach. One of these devices is the operational floating current conveyor (OFCC) [15], which exhibits flexible properties with respect to other current-mode or voltage-mode devices [16].

In order to fully exploit the current-mode differential scheme, a readout circuit has been designed based on the OFCC. The principal advantages of using OFCC as a basic unit in the readout circuit are as follows:

- It uses only one type of active elements (i.e., OFCC).
- The output of the readout circuit is a current instead of voltage, which is more suitable for subsequent A/D conversion.

11.2 Differential ISFET-Based pH Sensor

A differential ISFET consists of an ISFET and a reference FET (REFET) that have the same ISFET characteristics but are insensitive to the pH variation. Both are incorporated in the same device structure. A Si_3N_4/SiO_2 double layer is employed as ISFET gate insulator, whereas a buffered hydrogel layer forms the REFET insensitive membrane.

11.2.1 ISFET-Based pH Sensor

The operation of the ISFET pH sensor can be best explained by comparing it to that of a conventional MOSFET device. The ISFET can be viewed as a MOSFET in which the gate metal is replaced by an electrolytic solution in which a reference electrode is immersed, as shown in Figure 11.1(a). Figure 11.1(b) shows the schematic circuit representation of the ISFET. In particular, the interface between the metal gate and insulator is to be replaced by the reference electrode, the solution under test, and the pH sensitive insulating layer. Thus, the ISFET threshold voltage (V_T^*) can be deduced from the MOSFET (V_T) as shown in (11.1) [17]:

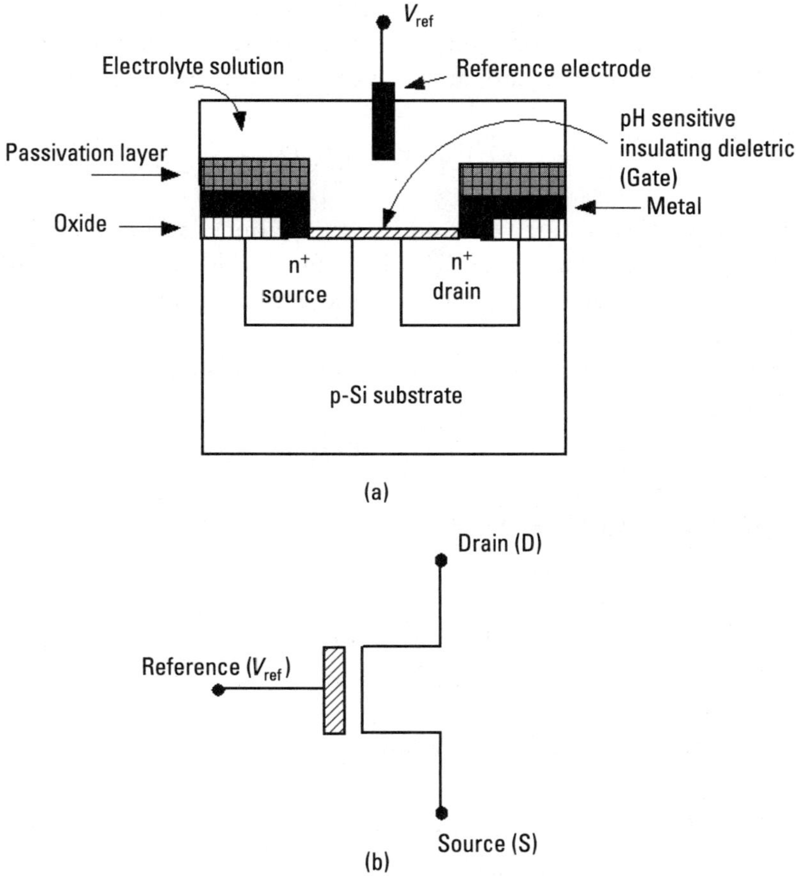

Figure 11.1 (a, b) Schematic cross-sectional view and circuit representation of the discrete ISFET [21].

$$V_T^* = V_T + E_{ref} + \varphi_{ij} + \chi_e - (\varphi_m / q) - \psi_o(\text{pH}) \tag{11.1}$$

where E_{ref} is the absolute reference electrode potential, φ_{ij} is the liquid junction potential difference between the reference solution and the electrolyte under test, χ_e is the electrolyte-insulator surface dipole potential, q is the electron charge, (φ_m/q) is the work function of the gate metal relative to vacuum, and ψ_o(pH) represents the potential difference between the insulator surface exposed to the electrolyte and the bulk of the electrolyte itself. This is the sole pH-dependent term in the ISFET threshold voltage. All the quantities on the right side of (11.1), except ψ_o (pH), are insensitive to changes in the solution pH. Thus, the threshold voltage (V_T) can be expressed as:

$$V_T^* = K_1 - \psi_o(\text{pH}) \tag{11.2}$$

where K_1 summarizes all the pH-independent quantities in (11.1).

The ISFET device is commonly biased to operate in the linear mode where $V_{DS} << V_{GS} - V_T^*$. The drain current in this case is given as:

$$I_{DS} = K\left[\left(V_{GS} - V_T^*\right) - \left(V_{DS}/2\right)\right]V_{DS} \tag{11.3}$$

where $K = \mu_n C_i W/L$, μ_n is the electron (for an n-channel device) mobility, C_i is the gate capacitance per unit area, W is the channel width, and L is the channel length. As for the voltages, V_{GS} is the gate-to-source voltage and V_{DS} is the drain-to-source voltage.

For $V_{DS}^2/2 < V_{GS} - V_T^*$, (11.3) can be approximated as:

$$I_{DS} \approx K\left[\left(V_{GS} - V_T^*\right)\right]V_{DS} \tag{11.4}$$

From (11.4), V_{GS} can be obtained as:

$$V_{GS} = \frac{I_{DS}}{KV_{DS}} + V_T^* \tag{11.5}$$

Combining (11.2) and (11.5) yields:

$$V_{GS} = \frac{I_{DS}}{KV_{DS}} + K_1 - \psi_o(\text{pH}) \tag{11.6}$$

For I_{DS} and V_{DS} held constant, V_{GS} can be written as:

$$V_{GS} = \text{Const} - \psi_o(\text{pH}) \tag{11.7}$$

It is evident from (11.7) that any change in the electrolyte pH will result in a corresponding change in the gate voltage V_{GS}.

11.2.2 Differential ISFET Sensor

Figure 11.2 is a schematic representation of the differential measurement setup. The differential ISFET measurement configuration was applied to effectively reduce any additional common mode disturbances such as temperature dependency and common-mode noise compensation [7]. This configuration also solved the reference electrode problem (the problem of the liquid junction potential, which can also react very quickly to the change in ionic composition, this po-

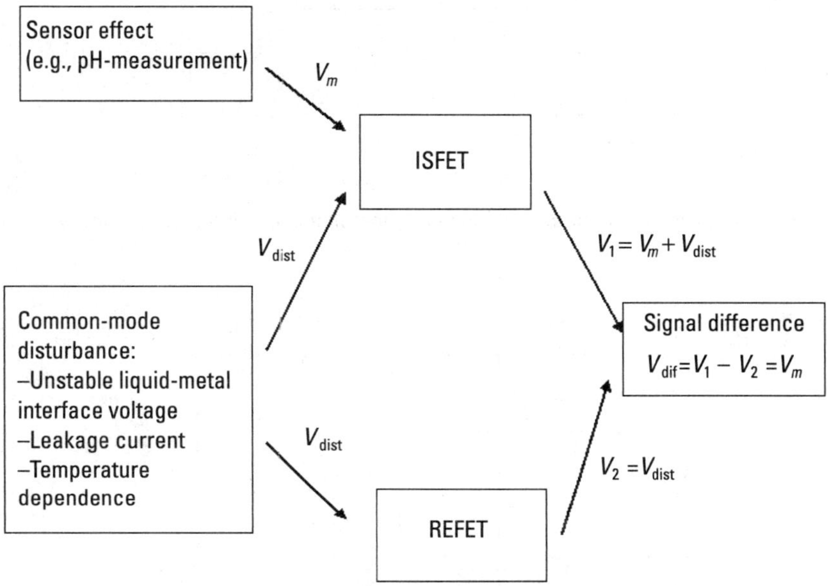

Figure 11.2 Differential measurement setup of an ISFET-system with a reference electrode [10].

tential change possibly coinciding with the ISFET response to the sample). The unstable electrode-reference electrode potential is a common signal to the differential system and will thus not interfere with a final output signal [19]. Figures 11.3 and 11.4 show both the block diagram and the schematic representation of the OFCC (refer to Chapter 10 for detailed information about OFCC).

11.3 pH Readout Circuit Based on an Operational Floating Current Conveyor

A schematic diagram of a readout circuit based on OFCC is shown in Figure 11.5. It provides an output current proportional to the difference in the voltage signals obtained from the ISFET and the REFET sensors. As shown in Figure 11.5, the readout circuit consists of two identical parts, each respectively connected to the ISFET sensor and the REFET sensor. Each part consists of three OFCCs. For the first part, OFCC (1) and OFCC (2) are configured as voltage followers, while OFCC (3) is a voltage to a current converter (i.e., a transconductance amplifier). Similarly, in the second part, OFCC (4) and OFCC (5) are configured as voltage followers, while OFCC (6) is a voltage-to-current converter.

The readout circuit provides the appropriate bias for the ISFET and the REFET sensors. OFCC (1), OFCC (2), OFCC (3), and OFCC (4), which are

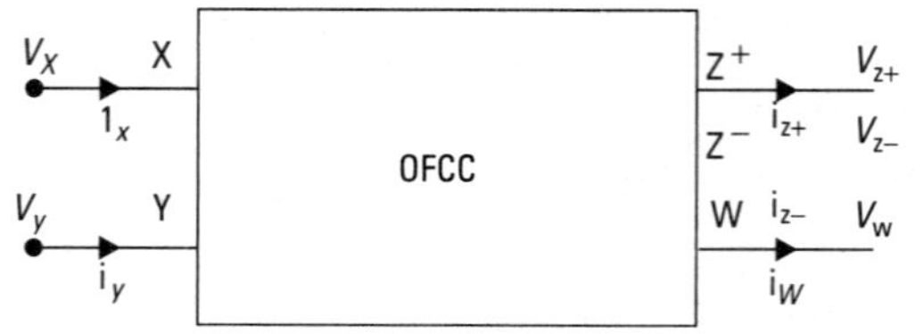

Figure 11.3 Block diagram representation of the operational floating current conveyor.

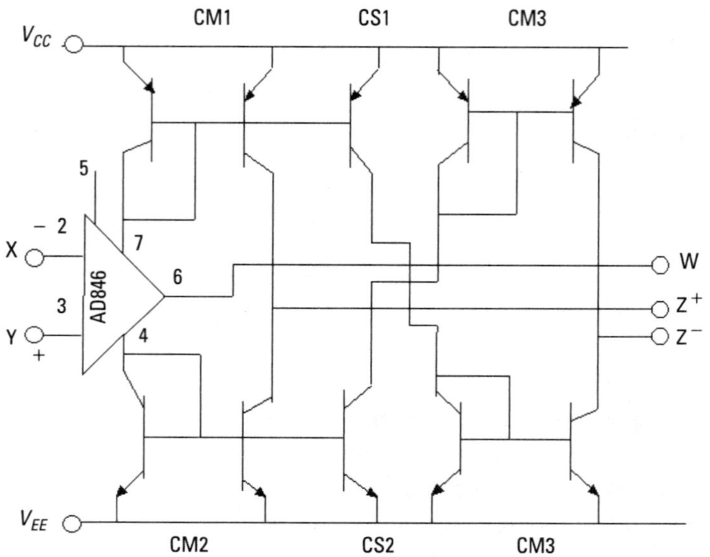

Figure 11.4 Circuit scheme of the OFCC.

configured as voltage followers, convey the voltage drop across R_1 and R_2 to the drain sources of the ISFET and REFET, respectively. To ensure good system accuracy, the circuit has to maintain the drain current and the drain-to-source voltage of both the ISFET and REFET devices as constantly as possible and independently of the threshold voltage. In this configuration, the drain current (I_{DS1}) of the ISFET is fixed by the current sink I_{S2} and the drain-to-source voltage (V_{DS1}) is fixed by the current source I_{S1} and the resistor R_1 as shown in Figure 11.5. Similarly, the drain current (I_{DS2}) of the REFET is fixed by the current sink I_{S4} and V_{DS2} is fixed by the current source I_{S3} and the resistor R_2 (see Figure 11.5). Any change in the pH of the solution results in a corresponding variation in the threshold voltage of both the ISFET and the REFET sensors. This in turn influences V_{GS}; see (11.7). Now since V_G is the reference voltage and can take any value, in our circuit $V_G = 0$, the change in pH will change V_A and V_B. As we mentioned previously, the ISFET is more sensitive to the variation of the pH than the REFET. Therefore, V_A will be higher than V_B. V_A and V_B are conveyed

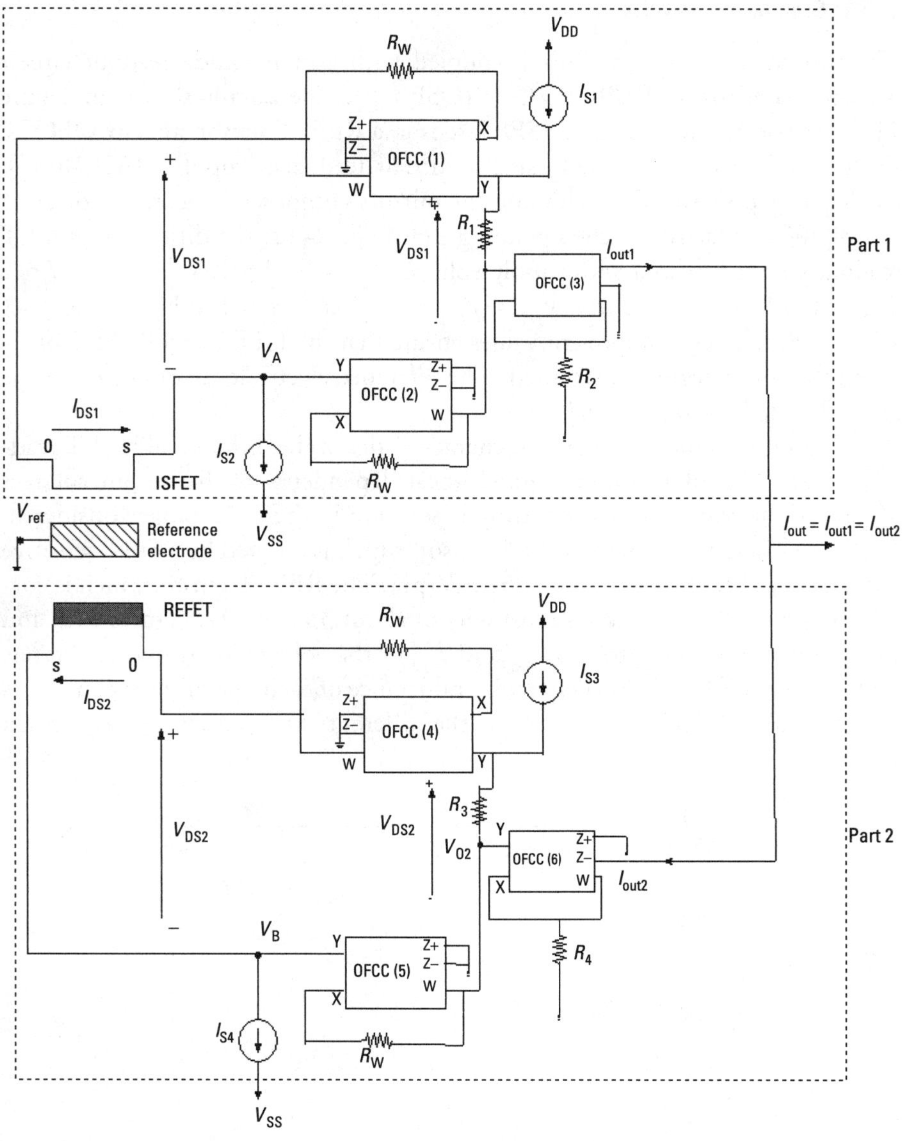

Figure 11.5 New differential ISFET current mode readout circuit (*From:* [21]. Reprinted with permission.)

to the output terminals V_{o1} and V_{o2} via OFCC (2) and OFCC (5), respectively. In order to convert V_{o1} and V_{o2}, which represent the change in pH into current (i.e., I_{out1} and I_{out2}), OFCC (3) and OFCC (6) are used. The total output current (I_{out}) is the difference between I_{out1} and I_{out2} (i.e., $I_{out} = I_{out1} - I_{out2}$). Since I_{out1} is related to the output voltage coming from the ISFET (V_{out1}) and I_{out2} is related to the output voltage coming from the REFET (V_{out2}), I_{out} is related to the deferential voltage $V_{out1} - V_{out2}$.

11.3.1 Simulation Results

The proposed differential ISFET coupled with current-mode readout circuit was simulated using PSPICE-2G (BIOSPICE). The circuit shown in Figure 11.5 was constructed using an ISFET sensor, a REFET sensor, and six OFCCs. Each OFCC was constructed using a current feedback amplifier (AD846AQ, Analog Devices), together with current mirrors composed of a transistor array (CA3096CE, Harris). In the operating pH range 2–12, the differential ISFET readout circuit is biased with supply voltages $V_{DD} = |V_{SS}| = 10\text{V}$, $I_{S2} = I_{S4} = I_{DS1} = I_{DS2} = 50\ \mu\text{A}$, $I_{S1} = I_{S3} = 2\ \mu\text{A}$, $V_{DS1} = V_{DS2} = 0.2\text{V}$, $R_1 = R_3 = 100\ \text{k}\Omega$ and $R_W = R_2 = R_4 = 1\ \text{K}\Omega$. These component values ensure that the ISFET and REFET operate in the linear region as required. The reference electrode is connected to the ground (i.e., $V_{ref} = 0$ potential).

Figure 11.6 shows the output characteristics of the differential ISFET readout circuit. This plot exhibits a good linear dependence of the output voltages V_{o1} and V_{o2} on the pH of the electrolytic solution in the 2–12 range considered. As shown in Figure 11.6, the ISFET sensor, which is covered by a Si_4N_3 sensitive layer, has a pH sensitivity of about 52 mV/pH. The REFET sensor, which is covered by a SiO_2 layer, has a pH sensitivity of about 33 mV/pH. Figures 11.7 and 11.8 show the output currents I_{out1} and I_{out2} of the voltage to current converters OFCC (3) and OFCC (6). The transconductance of each one of these converters is 0.093 mA/0.1V. Figure 11.9 shows the differential output current I_{out}, which

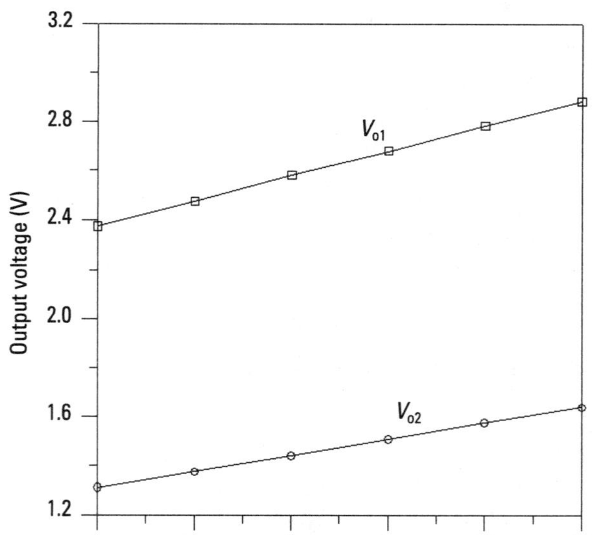

Figure 11.6 Plot of the output voltage V_{o1} and V_{o2} versus solution pH. (*From:* [21]. Reprinted with permission.)

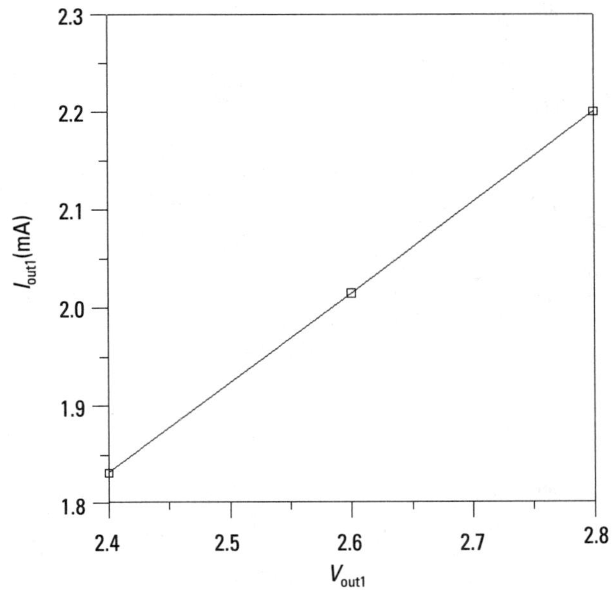

Figure 11.7 Plot of the output current (I_{out1}) versus V_{out1}. (*From:* [21]. Reprinted with permission.)

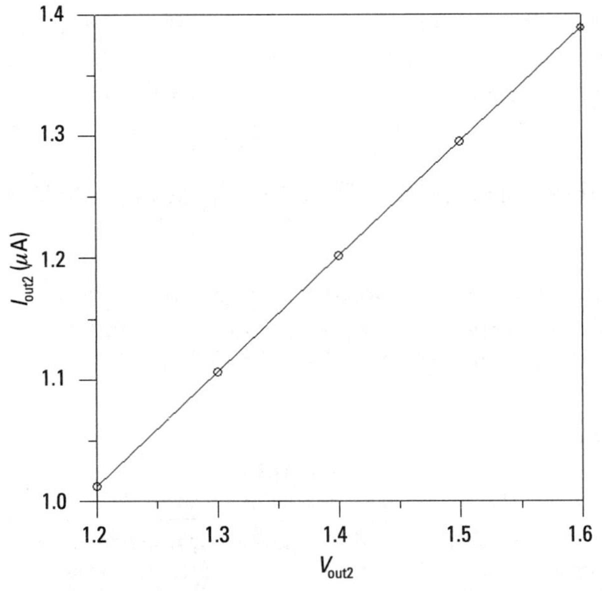

Figure 11.8 Plot of the output current (I_{out2}) versus V_{out2}. (*From:* [21]. Reprinted with permission.)

represents the difference between I_{out1} and I_{out2} with respect to the differential voltage $V_{out1} - V_{out2}$. The simulation results demonstrate that the proposed current mode readout circuit works properly and provides a good performance.

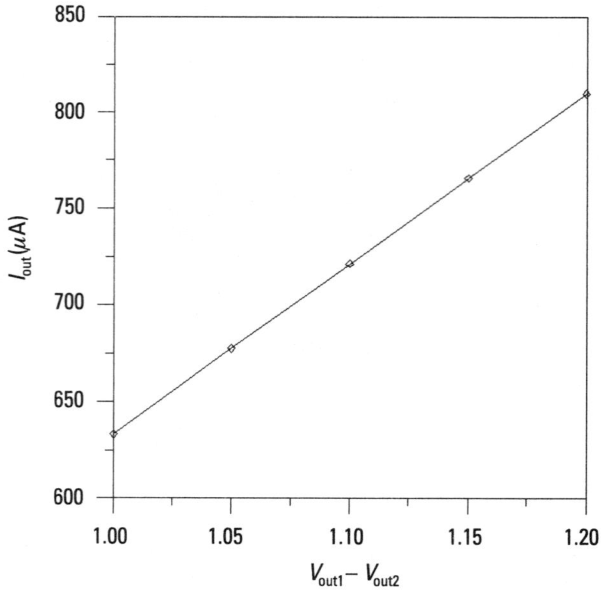

Figure 11.9 Plot of the output current (I_{out}) versus the differential output voltage ($V_{out1}-V_{out2}$). (*From:* [21]. Reprinted with permission.)

Table 11.1 shows a comparison between different read-out circuits with respect to constructions, the pH sensitivity, the type of the output (i.e., voltage or current), and the number of active elements used to implement these circuits.

11.4 pH Readout Circuit Using Only Two Operational Floating Current Conveyors

A schematic diagram of the proposed readout circuit is shown in Figure 11.10 [20]. It provides an output current proportional to the difference in the current signals obtained from the ISFET and the REFET sensors. As shown in Figure

Table 11.1
Comparison Between Different ISFET-Based pH Sensor Readout Circuits

Reference	Sensitive Layer Used	pH Sensitivity mV/pH	Type of Active Elements Used	Kind of Output
[9]	Si_3N_4	52	2	Current
[11]	SnO_2	58	1 (op-amp)	Voltage
[13]	Si_3N_4	58	1 (op-amp)	Voltage
This chapter	Si_3N_4	52	1	Current

11.10, the readout circuit consists of two OFCC, each respectively connected to the ISFET sensor and the REFET sensor. OFCC (1) and OFCC (2) are conveying the current pass through ISFET and REFET, respectively, to the output terminals (i.e., I_1 and I_2; see Figure 11.10), and then these currents are subtracted at the output node and pass into the load resistance (R_L). Therefore, in the case of $I_{DS1} = I_{DS2}$, there is no current difference at the output ($I_{out} = 0$). Thus, we can get a high common-mode rejection (CMR). In this configuration, the drain-to-source voltage (V_{DS1}) of the ISFET is fixed by connecting a voltage source V_{DD} to the drain. As a result of the voltage tracking between the X and Y terminals of the OFCC, the source voltage $V_{s1} = V_{y1} = 0$, so the source voltage is fixed, and the gate voltage (the reference electrode) is fixed by connecting a battery to the reference electrode (see Figure 11.4). Therefore, the conditions mentioned in (11.4) are now valid, and any change in the pH of the solution results in a corresponding variation in the threshold voltage of the ISFET sensor. This in turn influences the output current of the ISFET (I_{DS1}). Similarly, the drain voltage (V_{DS2}) of the REFET is fixed by the voltage source V_{DD}, the gate voltage is constant (see Figure 11.4), any change in the pH of the solution results in a corresponding variation in the threshold voltage of the REFET sensor, and consequently I_{DS2} will change as well [see (11.4)]. As we mentioned previously, the ISFET is more sensitive to the variation of the pH than the REFET. Therefore, I_{DS1} will be higher than I_{DS2}. I_{DS1} and I_{DS2} are conveyed to the output terminal via OFCC (1) and OFCC (2),

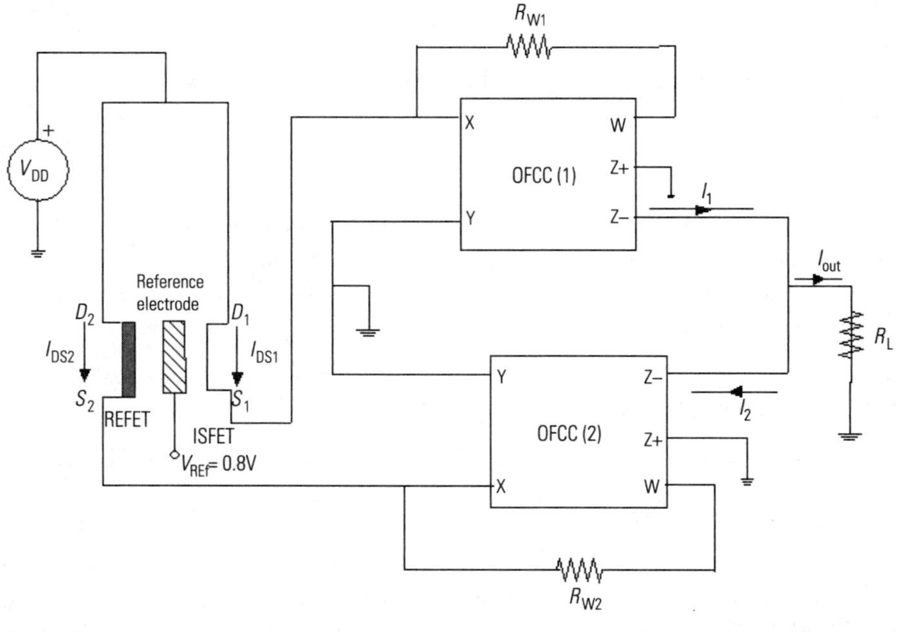

Figure 11.10 Current-mode readout circuit using 2-OFCC. (Reprinted from [20] with permission.)

respectively. The total output current (I_{out}) is the difference between I_{DS1} and I_{DS2} (i.e., $I_{out} = I_{DS1} - I_{DS2}$).

Taking into consideration the current tracking errors of the OFCC, the current tracking error between ports *X*, *Z*+, and *Z*– is:

$$\alpha = 1 - \varepsilon_{+} \tag{11.8}$$

and

$$\gamma = 1 - \varepsilon_{-} \tag{11.9}$$

where ε_{+} and ε_{-} denote the finite current tracking error at the high impedance outputs *Z*+ and *Z*–, respectively. Thus, port currents may then be expressed as $I_{Z+} = \alpha Ix$ and $I_{Z-} = \gamma Ix$. Therefore, I_1, and I_2 are defined by (11.7) and (11.8), respectively:

$$I_1 = \alpha_1 I_{DS1} \tag{11.10}$$

$$I_2 = \gamma_2 I_{DS2} \tag{11.11}$$

Therefore, the output current I_{out}, which is the difference of I_{DS1} and I_{DS2}, can be obtained from (11.10) and (11.11):

$$I_{out} = I_1 - I_2 = \alpha_1 I_{DS1} - \gamma_2 I_{DS2} \tag{11.12}$$

For ideal OFCCs, $\alpha_1 = \gamma_2 = 1$. Thus, the output current is:

$$I_{out} = I_1 - I_2 \tag{11.13}$$

From (11.13), we can observe that the output current is the difference between I_{DS1} and I_{DS2}, as required. In other words, we can observe that the proposed topology has a high common-mode rejection (CMR) for the common-mode current (i.e., $I_1 = I_2$).

11.4.1 Simulation Results

The OFCC has been constructed using a current feedback amplifier (AD846AQ, Analog Devices) together with the current mirrors composed of transistor arrays (CA3096CE, Harris). The differential ISFET coupled with current-mode readout circuit was simulated, using PSPICE-2G (BIOSPICE). The circuit shown in Figure 11.10 was constructed using an ISFET sensor, a REFET sensor and two OFCCs. In the operating pH range 2–12, the differential ISFET readout circuit is biased with supply voltages $V_{DD} = V_{DS1} = V_{DS2} = 0.2V$, and $R_{W1} = R_{W2} = R_L = 1$ KΩ. The reference electrode is connected to a constant voltage (i.e., $V_{ref} = 0.8V$). These values ensure that the ISFET and REFET operate in the linear region as required. Figure 11.11 shows the output characteristics of the differ-

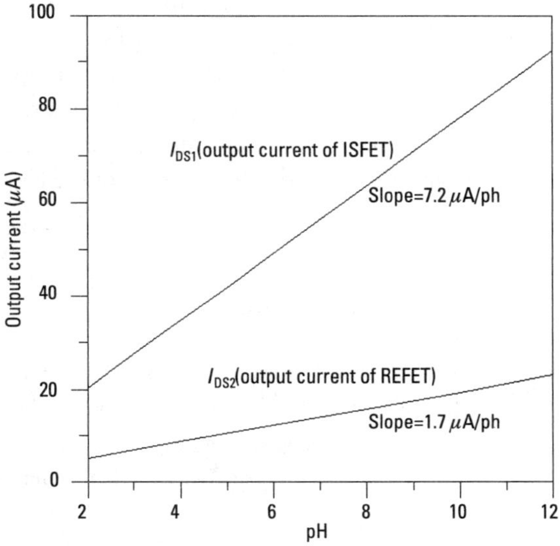

Figure 11.11 The output currents of the ISFET and REFET. (Reprinted from [20] with permission.)

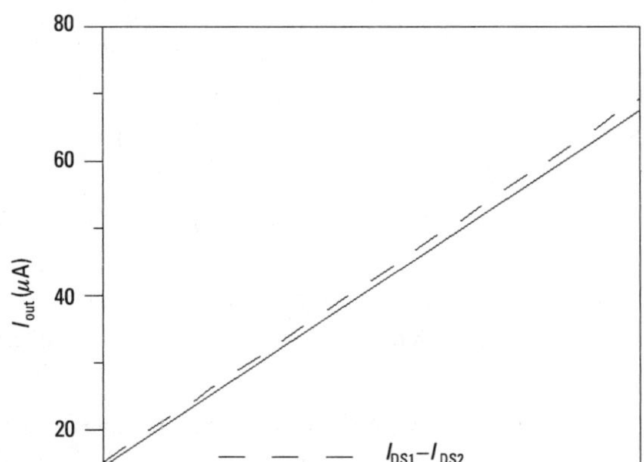

Figure 11.12 The output current I_{out} versus the difference current between I_{DS1} and I_{DS2}. (Reprinted from [20] with permission.)

ential ISFET readout circuit. This plot exhibits a good linear dependence of the output currents I_{DS1} and I_{DS2} on the pH of the electrolytic solution in the 2–12 range considered. As shown in Figure 11.11, the ISFET sensor, which is covered by a Si_4N_3 sensitive layer, has a pH sensitivity of about 7.2 μA/pH. The REFET sensor, which is covered by a SiO_2 layer, has a pH sensitivity of about 1.7 μA/pH. Figure 11.12 shows the differential output current I_{out}, which represents the difference between I_{DS1} and I_{DS2} with respect to the pH. The simulation results demonstrate that the proposed current-mode readout circuit works properly and provides a good performance. The small discrepancy between I_{out} and $I_{DS1}-I_{DS2}$ is due to the nonidealities of the active components [i.e., the current tracking errors between *W* and *Z*+ and the *W* and *Z*– terminals of OFCC (1) and OFCC (2), respectively].

References

[1] Begveld, P., "Development of an Ion-Sensitive Solid State Device for Neuropsychological Measurements," *IEEE Trans. on Biomedical Engineering*, Vol. BE-17, 1970, pp. 70–71.

[2] Lui, A., B. Margesin, and M. Zen, "Chemical Sensors Based on ISFET Transducers," *International Conf. Symposium on Devices and Materials*, Nova Gorica, Slovenia, 1996, pp. 51–72.

[3] Dzahini, K., F. Gaffiot, and M. Le Helley, "Using CMOS ASIC Technology for the Development of an Integrated ISFET Sensor," *Euro ASIC '91*, Paris, France, 1991, pp. 356–359.

[4] Cane, C., et al., "Compatibility of ISFET and CMOS Technologies for Smart Sensors," *TRANSDUCERS '91*, San Francisco, CA, 1991, pp. 225–228.

[5] Gimmel, P., et al., "Microstructure Solid-State Ion Sensitive Membranes by Thermal Oxidation of Ta," *Sensors and Actuators*, 1990, pp. 354–349.

[6] Gimmel, P., K. D. Schierbaum, and W. Gopel, "Reduce Light Sensitivity in Optimized Ta_2O_5 ISFET Structures," *Sensors and Actuators*, 1991, pp. 135–140.

[7] Janata, J., and R. J. Huber, *Solid State Chemical Sensor*, San Diego, CA: Academic Press, 1986.

[8] Wong, H. S., and M. H. White, "A CMOS Integrated ISFET-Operational Amplifier Chemical Sensor Employing Differential Sensing," *IEEE Trans. on Electron Devices*, Vol. 36, No. 3, 1989.

[9] Palan, B., E. Santos, and J. Karam, "A New ISFET Sensor Interface Circuit," *Proc. of the 1998 Eurosensors*, Southampton, United Kingdom, 1998, pp. 1–3.

[10] Muller, E., et al., "Differential ISFET/REFET as a Reference System for Integrated ISFET-Sensor Arrays," *International Conference on Solid-State Transducers '91*, 1991, pp. 467–470.

[11] Ravezzi, L., et al., "A CMOS ASIC for Differential Read-Out of ISFET Sensor," *ICECS 2001*, Malta, 2001, pp. 1513–1516.

[12] Chin, Y., et al., "A Novel SnO_2/Al Discrete Gate ISFET pH Sensor with CMOS Standard Process," *Sensors and Actuators*, Vol. B75, 2001, pp. 36–42.

[13] Ravezzi, L., and P. Conci, "ISFET Sensor Coupled with CMOS Read-Out Circuit Microsystem," *Electronics Letters*, Vol. 341, 1998, pp. 2234–2235.

[14] Finvers, I. G., et al., "On the Design of CMOS Current Conveyors," *Can. J. Elect. & Comp. Eng.*, Vol. 26, No. 1, January 2001.

[15] Khan, A., M. Al-Turiaia, and M. Abo El-Ela, "Operational Floating Current Conveyor: Characteristics, Modeling and Applications," *IEEE IMTC94*, Hamamtsu, Japan, 1994, pp. 788–790.

[16] Ghallab, Y. H., M. Abo El-Ela, and M. Elsaid, "Operational Floating Current Conveyor: Characteristics, Modeling and Experimental Results," *Proc. of the International Conference on Microelectronics, ICM99*, Kuwait, 1999.

[17] Martinoia, S., G. Massobrio, and M. Grattarola, "Modeling H + -Sensitive with SPICE," *IEEE Trans. on Electron Devices*, Vol. 39, No. 4, 1992, pp. 813–819.

[18] Bergveld, P., "Future Applications of ISFETs," *Sensors and Actuators*, Vol. B4, 1991, pp. 125–133.

[19] Soclof, S., *Design and Applications of Analog Integrated Circuits*, Englewood Cliffs, NJ: Prentice-Hall, Ch. 9, 1991, pp. 443–460.

[20] Ghallab, Y. H., and W. Badawy, "A Novel pH Sensor Current Mode Read-Out Circuit Using Only Two Operational Floating Current Conveyor," *Biomedical Circuit and Systems (BioCAS 2004)*, Singapore, December 1–3, 2004, pp. S1.5-13–S1.5-16.

[21] Ghallab, Y. H., and W. Badawy, "A Novel pH Sensor Current Mode Read-Out Circuit Using Operational Floating Current Conveyor," *ICMENS 2004*, Banff, Canada, 2004, pp. 262–265.

List of Symbols

Operators

∇ Gradient (del) operator

$|..|$ Absolute value

Re[..] Real part of complex number

Im[..] Imaginary part of complex number

f_o Angular frequency

Symbols

ε_o Permittivity of free space (8.854×10^{-12} F/m)

P Dipole moment

P_{eff} Effective dipole moment

a	Particle radius
K	Claussius-Mosotti polarization coefficient
σ_p	Particle conductivity
σ_m	Medium conductivity
τ_{MW}	Maxwell-Wagner charge relaxation time
ω_c	Crossing frequency
ω	Angular frequency
λ	Line charge density
z	Vertical position in frame of reference
ρ	Radial position in cylindrical coordinates
h	Height above the ground plane
Φ_p	Potential in spherical coordinate
$\mathbf{F}$	Force
ε_p	Particle permittivity
ε_m	Media permittivity
E	Electric field strength
V	Scalar electrical potential
I	Electric current
j	$\sqrt{-1}$
q	Free charge
K_n	Polarization coefficient

K_2 Quadrupole r^5 polarization coefficient

Z Impedance

R Resistance

About the Authors

Yehya H. Ghallab is an assistant professor at the Department of Biomedical Engineering, Helwan University, Cairo, Egypt. He is currently an engineer with SNC Lavailn T&D and a research associate with the Department of Electrical and Computer Engineering, University of Calgary. Dr. Ghallab is a holder of three patents in the United States and Canada, and he has authored and coauthored more than 30 papers. His research interests include the application of microelectronics in biomedicine, dielectrophoresis, microelectronic biomanipulators, sensors, and current-mode devices. Dr. Ghallab is a technical reviewer for several IEEE journals and conferences. He is an IEEE member and a professional engineer in Alberta and British Columbia, Canada. Dr. Ghallab received his B.Sc. and M.Sc. from the Electronics and Communication Department, Ain Shams University, Cairo, Egypt, in 1995 and 2000, respectively. He received his Ph.D. from the Electrical and Computer Engineering Department, University of Calgary, Calgary, AB, Canada, in 2005.

Wael Badawy is the founder and CEO of IntelliView Inc. He received his B.Sc. and M.Sc. from the Department of Computer Science and Automatic Control Engineering, University of Alexandria, Egypt, and his M.S. and Ph.D. from the Center for Advanced Computer Studies, University of Louisiana, Lafayette. In 2000, he joined the Department of Electrical and Computer Engineering, University of Calgary, Calgary, AB, Canada. Dr. Badawy has authored and coauthored more than 80 refereed journal/conference papers and approximately 30 technical reports. Dr. Badawy is a holder of six patents in the United States and

Canada. He is the guest editor for the special issue on systems-on-a-chip for real-time applications of the *Canadian Journal on Electrical and Computer Engineering*. Dr. Badawy was the technical chair for the 2004 International Workshop on SoC for Real-Time Applications, and a technical reviewer for several IEEE journals and conferences. He is currently a member of the IEEE-CAS Technical Committee on BioCas, VLSI, and Communications. He has been awarded the 2002 Petro Canada Young Innovator Award, the 2001 Micralyne Microsystems Design Award," and the 1998 Upsilon Pi Epsilon Honor Society and IEEE Computer Society Award for Academic Excellence in Computer Disciplines.

Index

Recent Titles in the Artech House Integrated Microsystems Series

Acoustic Wave and Electromechanical Resonators: Concept to Key Applications, Humberto Campanella

Adaptive Cooling of Integrated Circuits Using Digital Microfluidics, Philip Y. Paik, Krishnendu Chakrabarty, and Vamsee K. Pamula

Fundamentals and Applications of Microfluidics, Second Edition, Nam-Trung Nguyen and Steven T. Wereley

Integrated Interconnect Technologies for 3D Nanoelectronic Systems, Muhannad S. Bakir and James D. Meindl, editors

Introduction to Microelectromechanical (MEM) Microwave Systems, Héctor J. De Los Santos

An Introduction to Microelectromechanical Systems Engineering, Nadim Maluf

Lab-on-a-Chip: Techniques, Circuits, and Biomedical Applications, Yehya H. Ghallab and Wael Badawy

MEMS Mechanical Sensors, Stephen Beeby et al.

Micro and Nano Manipulations for Biomedical Applications, Tachung C. Yihllie Talpasanu

Microfabrication for Microfluidics, Sang-Joon John Lee and Narayan Sundararajan

Microfluidics for Biotechnology, Second Edition, Jean Berthier and Pascal Silberzan

Organic and Inorganic Nanostructures, Alexei Nabok

Post-Processing Techniques for Integrated MEMS, Sherif Sedky

Pressure-Driven Microfluidics, Václav Tesař

RFID-Enabled Sensor Design and Applicatons, Amin Rida, Li Yang, and Manos Tentzeris

RF MEMS Circuit Design for Wireless Communications, Héctor J. De Los Santos

Wafer-Level Testing and Test During Burn-in for Integrated Circuits, Sudarshan Bahukudumbi Krishnendu Chakrabarty

Wireless Sensor Network, Nirupama Bulusu and Sanjay Jha